Manufacturing Technology: Materials, Processes and Equipment

Manufacturing Technology: Materials, Processes and Equipment

Contributors

Don Ajith Rohana Dolage, Abu Bakar Sade et al.

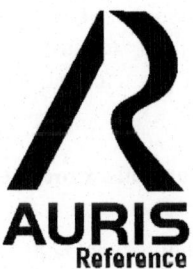

AURIS
Reference

www.aurisreference.com

Manufacturing Technology: Materials, Processes and Equipment

Contributors: Don Ajith Rohana Dolage, Abu Bakar Sade et al.

Published by Auris Reference Limited

www.aurisreference.com

United Kingdom

Manufacturing Technology: Materials, Processes and Equipment

ISBN: 978-1-78154-929-2
British Library Cataloguing in Publication Data
A CIP record for this book is available from the British Library

Printed in the United Kingdom

Exclusively distributed by CBS Publishers & Distributors Pvt. Ltd.

Sales & Distribution Rights only for India, Pakistan, Bangladesh, Sri Lanka, Nepal and Bhutan. This book is not to be sold outside these territories.

Contents

List of Abbreviations

ACGIH	American Conference of Governmental Industrial Hygienists
AIHA	American Industrial Hygiene Association
ANSI	American National Standards Institute
APF	assigned protection factor
ASHRAE	American Society of Heating, Refrigerating, and Air Conditioning Engineers
BSC	biological safety cabinet
BSI	British Standards Institute
CAV	constant air volume
CDC	Centers for Disease Control and Prevention
CNF	carbon nanofiber
CNT	carbon nanotube
CPC	condensation particle counter
CVD	chemical vapor deposition
DMPS	differential mobility particle sizer
ELPI	electrical low pressure impactor
EPA	Environmental Protection Agency
FFR	filtering facepiece respirator
FMPS	fast mobility particle sizer
HEPA	high efficiency particulate air
HSE	Health and Safety Executive
IH	industrial hygiene
LEV	local exhaust ventilation
LPM	liters per minute
MPPS	most penetrating particle size
MSDS	material safety data sheet
MUC	maximum use concentration
NIOSH	National Institute for Occupational Safety and Health
OEL	occupational exposure limit
PEL	permissible exposure limit
PHA	preliminary hazard assessment
PM	preventive maintenance
PPE	personal protective equipment
PtD	prevention through design
R&D	research and development
REL	recommended exposure limit
SMACNA	Sheet Metal and Air Conditioning Contractors' National Association
SMPS	scanning mobility particle sizer
SOP	standard operating procedures
TEM	transmission electron microscopy

TEOM	tapered element oscillating microbalance
TLV	threshold limit value
TWA	time- weighted average
VAV	variable air volume
VEM	video exposure monitoring

List of Contributors

Don Ajith Rohana Dolage
International Graduate School of Business, University of South Australia, Adelaide, Australia

Abu Bakar Sade
Faculty of Management, Multimedia University, Cyberjaya, Malaysia

Soosung Kim
Research Reactor Fuel Development Division, Korea Atomic Energy Research Institute, Yuseong-gu, Daejeon 305-353, Republic of Korea

Kihwan Kim
Research Reactor Fuel Development Division, Korea Atomic Energy Research Institute, Yuseong-gu, Daejeon 305-353, Republic of Korea

Jungwon Lee
Research Reactor Fuel Development Division, Korea Atomic Energy Research Institute, Yuseong-gu, Daejeon 305-353, Republic of Korea

Jinhyun Koh
Korea University of Technology and Education, Cheonan 330-708, Republic of Korea

Konstantin Chuntonov
NanoShell Consulting, Nitzanim, Migdal Haemek, Israel

Janez Setina
Institute of Metals and Technology, Ljubljana, Slovenia

Gary Douglass
Agile Chemistry, Inc., Elmhurst, USA

Vladimir Stepanovich Kondratenko
Moscow State University of Instrument Engineering and Computer Science (MGUPI), Moscow, Russia

Vladimir Evgenievich Borisovsky
Moscow State University of Instrument Engineering and Computer Science (MGUPI), Moscow, Russia

Alexandr Sergeevich Naumov
Moscow State University of Instrument Engineering and Computer Science (MGUPI), Moscow, Russia

Nikolay Eduardovich Petruljanis
Moscow State University of Instrument Engineering and Computer Science (MGUPI), Moscow, Russia

Kedar Mallik Mantrala
Department of Mechanical Engineering, Vasireddy Venkatadri Institute of Technology, Guntur, India

Mitun Das
Bioceramics and Coating Division, CSIR-Central Glass and Ceramic Research Institute (CGCRI), Kolkata, India

Krishna Balla
Bioceramics and Coating Division, CSIR-Central Glass and Ceramic Research Institute (CGCRI), Kolkata, India

Ch. Srinivasa Rao
Department of Mechanical Engineering, Andhra University College of Engineering, Visakhapatnam, India

V. V. S. Kesava Rao
Department of Mechanical Engineering, Andhra University College of Engineering, Visakhapatnam, India

Hasan Hosseini-Nasab
School of Engineering, Yazd University, P.O. Box 89195-741, Yazd, Iran

Mohammad Dehghani
School of Engineering, Yazd University, P.O. Box 89195-741, Yazd, Iran

Amin Hosseini-Nasab
School of Engineering, Yazd University, P.O. Box 89195-741, Yazd, Iran

J. Li
University Campus STeP Ri Slavka Krautzeka 83/A 51000 Rijeka, Croatia

J.Y H. Fuh
University of California, Los Angeles (UCLA), Los Angeles, California, United States.

Y.F. Zhang
National University of Singapore, Lower Kent Ridge Rd, Singapore

A.Y.C. Nee
University Campus STeP Ri Slavka Krautzeka 83/A 51000 Rijeka, Croatia

Vittorio Cesarotti
University of Rome "Tor Vergata", Italy

Alessio Giuiusa
University of Rome "Tor Vergata", Italy
Area Manager Inbound Operations at Amazon.com

Vito Introna
University of Rome "Tor Vergata", Italy

Dr. Tauseef Aized
Professor, Department of Mechanical, Mechatrnics and Manufacturing Engineering, KSK Campus, University of Engineering and Technology, Lahore, Pakistan

Tritos Laosirihongthong
University Campus STeP Ri Slavka Krautzeka 83/A 51000 Rijeka, Croatia

Anderson Vicente Borille
Technological Institute of Aeronautics - ITA Brazil

Jefferson de Oliveira Gomes
Technological Institute of Aeronautics - ITA Brazil

Preface

Manufacturing technology provides the productive tools that power a growing, stable economy and a rising standard of living. The text *Manufacturing Technology: Materials, Processes and Equipment* introduces and elaborates on the field of manufacturing technology—its processes, materials, tooling, and equipment. It covers the basics as well as the most recent advances in manufacturing technology. First chapter examines the impact of the adoption of flexible manufacturing technology (FMT) on the technical efficiency of Malaysia manufacturing industry. In second chapter, we focus on nanomaterial production and downstream handling processes. Third chapter presents an outline of the developed welding equipment for nuclear fuel bundle fabrication and reviews a conceptual design of remote welding equipment using a master-slave manipulator. The objective of fourth chapter is to provide application specialists with the first knowledge about how the new technologies work and what can be expected from them. New technology for grids and scales manufacturing in optical devices has been introduced in fifth chapter. The aim of sixth chapter is to understand the influence of post-fabrication heat treatment on microstructure, hardness, wear, and corrosion properties of laser deposited Co-Cr-Mo alloy. In seventh chapter, a dynamic model is presented in which lean manufacturing is linked with technology by causal relationships. The objective of eighth chapter is to develop a distributed collaborative design environment for supporting cooperation among existing engineering tools organized as independent agents on different platforms. In ninth chapter, we show how overall equipment effectiveness (OEE) can be used to carry out a correct equipments sizing and an effective production system design, taking into account both equipment time losses and their propagation throughout the whole production system. Tenth chapter deals with materials handling in flexible manufacturing systems. The purposes of eleventh chapter are to investigate manufacturing technology used in the Thai automotive industry and examine findings concerning certain manufacturing technology dimensions. The aim of last chapter is to present different decision making approaches to choose an adequate RP process.

Chapter 1

A FRONTIER APPROACH TO MEASURING IMPACT OF ADOPTION OF FLEXIBLE MANUFACTURING TECHNOLOGY ON TECHNICAL EFFICIENCY OF MALAYSIAN MANUFACTURING INDUSTRY

Don Ajith Rohana Dolage[1], Abu Bakar Sade[2]

[1]International Graduate School of Business, University of South Australia, Adelaide, Australia

[2]Faculty of Management, Multimedia University, Cyberjaya, Malaysia

ABSTRACT

This paper examines the impact of the adoption of Flexible Manufacturing Technology (FMT) on the Technical Efficiency of Malaysia Manufacturing Industry. Owing to the potential multicollinearity, the Principal Component Analysis has been adopted to extract the most appropriate underlying dimensions of FMT in an effort to substitute the eight FMT variables. The study has been conducted within FMT intensively adopted 16 three-digit industries that encompass 50 five-digit industries covering the years 2000-2005. The results obtained from the two situations, one, including the industry fixed effects dummy variables and the other without these, are contrasted. It is found that the model that included the industry fixed effect dummy variables possesses a greater explanatory power. The two principal components that account for the greater variation in FMT show positive and moderately significant relationship with TE. The study concludes with sufficient evidence that FMT has a direct and moderately significant relationship with TE.

INTRODUCTION

Over the last two decades many researchers conducted studies that examined the factors attributing to the performance of the East Asian Economies. According to Mahadevan [1] in particular, the GDP growth of these economies was found to be driven by the perspiration factor of input accumulation rather than the inspiration factor of total productivity growth (TFPG). Owing to the increasing concern regarding the continued growth of these economies, new endogenous growth theories have been adopted in an effort to explain the performance of these economies. The significance of the manufacturing sector in these economies is so much that in each economy it accounts for well over 30 percent of the overall GDP. Therefore, more focused studies looking into the inspiration factor of TFPG, which also means factor productivity should be undertaken.

According to the conventional growth accounting methodology, TFPG is considered synonymous with technological progress (TP) which is also known as technical change. According to Coelli Rao, O'Donnell and Battese [2], TP represents advances made in technology that may be represented by an upward shift in the production frontier. This methodology which is a non frontier approach, is based on the assumption that all industries are fully realising their capacity in the production process and thus, are technically efficient. In this approach, no distinction is made between TP and changes in technical efficiency (TE) with which a known technology is adopted in production. As a result, this approach does not separately account for the technological improvement embodied in labour or the capital stock (change in efficiency) contained in the TFPG. Under a newer approach named Stochastic Frontier Production Function (SFPF), originally proposed by Aigner, Lovell and Schmidt [3], output growth is decomposed into input growth and TFPG and TFPG is in turn decomposed into TP and TE. The word frontier signifies the idea of maximality and represents the "best practice" approach to production. Mahadevan [4] showed that unlike the growth accounting approach which provided a shape of an average industry, the estimation of a frontier function was heavily influenced by the best performing industries. According to Mahadevan and Kalirajan [5], TE can be due to the accumulation of knowledge in the learning-by-doing process, improvements in the instructions for mixing together raw materials, diffusion of new technology, improved managerial practices or R&D undertaken by government or profit maximising agents, or can be affected by overall market structure of industry as it affects the methods used for acquiring, developing or modifying technology.

It is widely believed that intensive regimes of contemporary manufacturing paradigms such as mass customisation, customerisation and instant customerisation can pave the way for a competitive manufacturing industry. The studies show that mass customisation is the core manufacturing paradigm. The studies also showed that the crucial determinant of the successful implementation of mass customisation is the abundant use of Flexible Manufacturing Technology (FMT) [6,7]. Kumar and Desmukh [8] remark that today's customer not only expects quality, reliability and competitive pricing but also customised products with timely delivery, it is desirable that an organisation is as flexible as possible,. According to Sinha and Noble [9], FMT can represent a huge cost for adopting firms, but may also offer the chance to achieve competitive advantage through superior manufacturing. Therefore, it would be important to examine the causal link between the degree of FMT adoption and TE.

The average GDP growth of Malaysia during 2000- 2007 (5.5 percent) is lower than that during 1990-2000 (7.0 percent). Malaysian Manufacturing sector GDP during 2000-2007 (13.0 percent) is much lower than the same for the period 1990-2000 (4.8 percent). These are some of the key indicators to the declining competitiveness of the Malaysian manufacturing industry over the period 2000-2007. In the Malaysian manufacturing industry, FMT is widely adopted and has received the due attention from the industry policy makers. The Malaysian Industrial Development Authority (MIDA) [10] has recognised a number of promoted activities and products (for the development and production) with regard to high technology establishments. The engagement in these activities will make them entitled to pioneer status or investment tax allowance under the promotion of Investment Act 1986. This includes FMT products such as, Computer process control systems/equipment, Process instrumentation, and Robotic equipment and Computer numerical control machine tools. The Ninth Malaysia Plan which is aimed at achieving changes in the structure and improved performance of the economy with every economic sector achieving higher value added and total factor productivity. The "Thrust 1" of this Plan states that, "Application of high technology and production of higher value added products will be given emphasis. Measures will be undertaken to migrate the electrical and electronics (E&E) industry towards high-technology and higher value added activities". Hence, the Malaysian experience is used as a case study given its suitability.

Mahadevan [1] adopted SFPF in a study on TFPG of Malaysia's Manufacturing Industries. Sun and Kalirajan [11] found that 2.5 percent average annual rate of TP during this period was the major contributor to TFPG in the Korean manufacturing industry whereas TE grew by a modest 1.1 percent per annum. Zhang and Zhang [12] adopted SFPF to estimate the TE of China's large and medium sized iron and steel enterprises. Lee, Kim and Heo [13] in their empirical study on TP versus TE Gains in Manufacturing Sector of Korea, adopted the nonparametric Malmquist productivity index to break down the productivity growth into two components; technological change (innovation) and efficiency change (catching up). According to Battese and Broca [14], SFPF involves an unobservable random variable associated with the technical inefficiency of production of individuals, in addition to the random error in a traditional regression model. These studies show that TE is adopted extensively to measure efficiency of the manufacturing sector.

As shown above, evidently, only a few studies have examined the impact of specific technologies on the measures of competitiveness such as productivity, technical efficiency and profitability at industry level using less aggregated data. Berndt and Morrison [15] examined the impact of high-tech investments on multifactor productivity (MFP) and three profitability measures. While the study found only limited evidence of a positive relationship between profitability and the share of high-tech capital in the total physical capital stock, it establish that they were negatively correlated with MFP. Dolage, Sade and .Elsadig [16] investigated to impact of FMT on the TFPG of Malaysian manufacturing industry and established the significance of certain types of FMT on the TFPG.

Amato and Amato [17], Dolage and Sade [18] have investigated the impact of high-tech or FMT investments on Price Cost Margin. This study established that there was a positive impact from high-tech investments regardless of whether or not the specification includes industry effects dummy variables to account for the differences in technological opportunity among industries. Hence, according to productivity literature evidently no empirical study has been undertaken to investigate the impact of FMT adoption on TE of Malaysian manufacturing industry.

METHODOLOGY

The basic research hypothesis of the study is:

A high degree of FMT adoption enhances TE of the manufacturing industry of Malaysia.

Estimation of TE Using SFPF

The SFPF adopted in this study to compute the industrywise technical efficiency is based on the same adopted by Mahadevan [4] in the study on "A frontier approach to measuring TFPG in Singapore Manufacturing Industry". The derivation of the function to evaluate TE, based on SFPF is described below in Equation (1):

$$\ln Y_{it} = \delta_{1i} + \sum_{j=1}^{2} \delta_{ij} \ln W_{ijt}$$

(1)

where

$i = 1,2,3,\cdots,16$ (no of three-digit manufacturing industries);

j = K, and L (K-capital and L-labour);

t = 1,2,3 (no of years from 2000-2005);

Y = Value added output measured in 2000 prices;

W_K = Capital expenditure measured in 2000 prices;

W_L = Number of workers employed;

δ_{1i} =Intercept term of the ith three-digit manufacturing industry;

δ_{ij} = Actual response of output to the method of application of the jth input used by the ith manufacturing industry.

Mahadevan [19] in the study on "Is There a Real Growth Measure for Malaysia's Manufacturing Industries?" incorporated time dummies in the SFPF to capture the effects of time on the TE. Therefore, time dummies are incorporated in this study too and accordingly Equation (1) shown above can be modified to accommodate this effect; the revised model is represented in Equation (2) shown below:

$$\ln Y_{it} = \delta_{1i} + \sum_{j=1}^{2} \delta_{ij} \ln W_{ijt} + \sum_{t=1}^{6} \delta_t$$

(2)

Kalirajan and Shand [20] in their study on Frontier Production Functions and Technical Efficiency Measures, explained that the efficient use of inputs due to various industry specific characteristics contributed individually to the technical efficiency of the industry and the contributions can be measured by the magnitudes of the random slope coefficients. All other production characteristics are captured by the varying random intercept term.

Since intercepts and slope coefficients can vary across industries they can be represented as:

$$\delta_{ij} = \delta_j^1 + u_{ij}$$

(3)

$$\delta_{1i} = \delta_1^1 + v_{ij}$$

(4)

where δ_j^1 is the mean response coefficient of output with respect to the jth input, and u_{ij} and v_{1i} are random disturbance terms.

Combining Equations (2), (3) and (4), SFPF can be presented in Equation (5) shown below:

$$\ln Y_{it} = \delta_1^1 + \sum_{j=1}^{2} \delta_j \ln W_{ijt} + \sum_{j=1}^{2} u_{ij} \ln W_{ijt} + v_{1i} + \sum_{t=1}^{6} \delta_t$$

(5)

Mahadevan [21] showed that by adopting Aitken's generalised least squares method proposed by Hildredth and Houck and the estimation procedure by Griffith the industry-specific and input-specific response coefficient estimates of the above model could be obtained. The highest values of each response coefficient and the intercept determine the frontier coefficient of the potential production function. If δ^* denotes the parameter estimates of the frontier production function, then

$\delta_j^* = \max\{\delta_{ij}\}$ where i = 1, 2, ⋯16 and j = K, L

$\delta_1^* = \max\{\delta_{1i}\}$ where i = 1, 2, ⋯16

$\delta_t^* = \max\{\delta_t\}$ where t = 1, 2, ⋯ 6 Since $\delta_1^* = \delta_1^l + \max\{v_t\}$ $\delta_j^* = \delta_j^l + \max\{u_{ij}\}$, and j represents both K and L, the above equations can be rearranged as follows:

$\delta_K^* = \delta_K^l + \max\{u_{iK}\}$

$\delta_L^* = \delta_L^l + \max\{u_{iL}\}$

$\delta_1^* = \delta_1^l + \max\{v_t\}$

$\delta_t^* = \max\{\delta_t\}$

The potential output of each industry can be realised when each industry adopts the "best practice" techniques and the maximum potential output for each industry is given by Equation (6):

$$\ln Y_{it}^* = \delta_1^* + \sum_{j=1}^{2} \delta_j^* \ln W_{ijt} + \delta_t^*$$

(6)

In Equation (6), since j represent both capital (K) and labour (L) it can be expanded to form a new equation (Equation (7)) which is given below:

$$\ln Y_{it}^* = \delta_1^* + \delta_K^* \ln W_{iKt} + \delta_L^* \ln W_{iLt} + \delta_t^*$$

(7)

The industry-specific TE can be represented by Equation (8) shown below as the ratio of the industry's actual realised output to that of its potential output:

$$TE_{it} = \frac{Y_{it}}{\exp(\ln Y_{it}^*)}$$

(8)

The numerator Y_{it} can be computed simply as the "gross output" take away "cost of inputs", values for which were obtained from ASMI.

Equation (5) in which j is representative of both capital (K) and labour (L) can be expanded and presented as follows:

$$\ln Y_{it} = \delta_1^1 + \delta_K \ln W_{iKt} + \delta_L \ln W_{iLt} + u_{iK} \ln W_{iKt}$$
$$+ u_{iL} \ln W_{iLt} + v_{1i} + \sum_{t=1}^{6} \delta_t$$

(9)

All the values relating to the six years for the 50 MSIC five-digit industries (altogether 300 cases) are to be substituted in Equation (9) given above:

The technique of multiple regression analysis was adopted to ascertain the respective response coefficients for capital and labour and the constants (δ_1^*, δ_k^*, δ_L^* and δ_t^*). The widely used statistical software, Statistical Package for Social Sciences (SPSS) was adopted to run the above multiple regression and from its output, the maximum values for u_{iK}, u_{iL} and v_i were obtained to compute δ_1^*, δ_K^*, δ_L^* and δ_t^*. Once these values were ascertained, as the next step, they were substituted in the Equation (7) in order to compute the maximum frontier outputs for each industry for all the 6 years with the use of EXCEL.

Review of Factors Affecting TE in the Manufacturing Industry

Geographic Location (GLO): According to ASMI reports, Malaysia shows a significant diversity in the number of firms located within different provinces. Certain regions in a country could be disadvantaged by its distal location from the major markets due to less access to physical and human capital and technologies. Sun and Kalirajan [11] stated that firms applied their production technology and inputs differently as a result of the differences in location, experience, and firm size. Zhang and Zhang [12] stated that geographic location affected the efficiency of an enterprise. Vu [22], Margono and Sharma [23], Zhang and Zhang [12], Sun and Kalirajan [11] have considered GLO as an explanatory variable of TE.

Malaysia does not publish indicators which show the favourability of a particular province for an industry. The ASMI contains the number of establishments located within a province; usually Table 5 of ASMI depicts this information. The researcher assumes that the major reason why a large number of establishments are concentrated in a particular province is that it has a conducive environment for industries with respect to many aspects namely, access to skilled labour, technology, support industries, raw materials, close proximity to markets and ports etc. So, the corollary is that the number of firms in a province indicates the location advantage of the province. Accordingly, the provinces were ranked based on the number of establishments located in each. The directory published by Federation of Malaysian Manufacturers (FMM) contained the addresses of establishments including the province in which each establishment is situated [24]. In the FMM directory 2007, the manufacturing industries have been categorised under MSIC four-digit level. Thereafter, each establishment in a particular province was multiplied by its respective ranking. This ranking for a province was not a constant over the six year period considered since the number of establishments located within provinces has evidently changed yearly, though of course marginally. Finally, GLO was computed for each MSIC five-digit industry as the summation of the products of the number of establishment and the ranking of the respective province.

Ownership (OWN): It has long been established that generally foreign owned establishments are more efficient than locally owned establishments. Vu [22], Margono and Sharma [23], Zhang and Zhang [12], Sun and Kalirajan [11] and Mahadevan [25] considered OWN as an explanatory variable of TE. Since Malaysia is known to have a long history as a well considered recipient of FDI since mid seventies, the variable OWN was incorporated to account for the "ownership" impacts on TE.

Malaysia does not publish the status of establishments with respect to their ownership. Nevertheless, the DOS from its Economic Census data provided the number of establishments belonging to the Non Malaysian and Malaysian categories for each MSIC five-digit industry for all the years under review. Mahadevan [25] used a dummy variable "1" for industries in which more than 45 percent of the total number of establishments were either wholly foreign owned or joint ventures which were more than half foreign owned; otherwise "0". However, information provided by the DOS in this respect was not detailed enough to assess the degree of foreign ownership regarding the Non Malaysian firms. A dummy variable (OWN) was used to distinguish between industries that had higher and lower percentages of Non-Malaysian firms. In this study, dummy variable "1" was used if in a particular five-digit industry more than 30% of the establishments were Non-Malaysian otherwise, the dummy variable "0" was used.

Firm Age (FAGE): The length of existence of a firm can indicate advancement of different factors favourable to TE namely, propensity to employ skilled workers and degree of learning by doing. According to Zhang and Zhang [12], the maturity of a firm represented by its age is a factor contributing towards inefficiency but the firm size coefficients are expected to affect inefficiency negatively. Alternatively, vintage of capital can be used to detect technical progress in production, especially the type of technical progress embodied in capital (Zhang and Zhang [12]. However, measuring the vintage of capital is not straightforward owing to the lack of data on the past capital investments at the firm level. Since it is reasonable to assume that FAGE indicates the vintage of capital and the experience of the firm, it is considered as a control variable in the TE model of this study. Sun and Kalirajan [11] as well as Margono and Sharma [23] have considered FAGE as an explanatory variable of TE in their studies.

The DOS Malaysia does not publish data pertaining to FAGE. The FMM directory gives the year of incorporation of each establishment listed under MSIC four-digit level industry. Using this value, the average FAGE for each industry was computed as at year 2000. The average firm age (FAGE) for the later years was computed by increasing the average firm age as of 2000 by one for each subsequent year.

Incentive Payments (INC): It is rational to consider that incentive schemes and bonus payments can motivate employees to work hard resulting in efficiency gains for the industry. Mahadevan [25] highlights differential incentive systems as one of the factors causing inefficiency. However, in this study INC has not been incorporated in the TE model as an explanatory variable. Vu [22] has included bonus payments as one of the hypothesised influencing factors in the TE model adopted to study technical efficiency of industrial state-owned enterprises. In this study, INC was measured as the ratio of incentive payment to salary. The DOS Malaysia does not publish data for INC industry-wise. The researcher computed INC from the Economic Census data maintained by the DOS Malaysia.

Firm Size (FSIZE): This variable is a crude proxy for scale of entry barrier. Theoretically, the minimum efficient plant size is a better proxy but could not be included due to the non availability of data. Higher productivity gains can be expected in the presence of oligopolistic competition. Therefore, researchers include average plant size or firm size in TE models to take account of such effects: Chandrasiri [26], Amato and Amato [17] and McGuckin [27]. However, the direction of the link is ambiguous. In two separate studies, Amato and Amato [17] and Round [28] have incorporated FSIZE in their PCM model to account for the entry barriers. In this study, FSIZE was measured as the average firm size of the eight largest firms in each industry. As these data are not annually published in Malaysia, they had to be computed using the data obtained from the Economic Census data maintained by the DOS.

The eight types of FMT considered in this study are given below:

Computer Numerical Control Machine Tools (CNC): Measured as the percentage of firms in each MSIC fivedigit industry using microprocessor based numerical control technologies, referred to as computer numerical control machine tools. Numerical Controlled Machine Tools (NC): Measured as the percentage of firms in each MSIC five-digit industry using numerical controlled machine tools.

Robotics (ROB) Measured as the percentage of firms in each MSIC five-digit industry using robotics.

Programmable Logic Controllers (PLC): Measured as the percentage of firms in each MSIC five-digit industry using programmable logic controllers.

Automated Inspections (INS): Measured as the percentage of firms in each MSIC five-digit industry using automated sensor-based inspection, either during the production process or final product.

Automated Storage and Retrieval Systems (ASR): Measured as the percentage of firms in each MSIC fivedigit industry using automated storage and retrieval systems.

Computer Aided Design (CAD): Measured as the percentage of firms in each MSIC fivedigit industry using computer aided design to control manufacturing machinery.

Local Area Networks (LAN): Measured as the percentage of firms in each MSIC five-digit industry using local area networks.

Industry Fixed Effects Dummy Variables (IND_j): The study involved 50 five-digit industries included in 16 three-digit industries. It is logical to assume that industry characteristics among these 16 three-digit industries can be diverse and need to be captured by a variable. Therefore, 16 dummy variables (IND_j) were incorporated into the TE model to capture the industry fixed effects. The model representing the relationship among TE explanatory variables and FMT variables can be specified as given below:

Technical Efficiency Model

$$\begin{aligned} TE = {} & \lambda_0 + \lambda_1 GLO + \lambda_2 OWN + \lambda_3 FSIZE + \lambda_4 FAGE \\ & + \lambda_5 INC + \lambda_6 CNC + \lambda_7 NC + \lambda_8 ROB + \lambda_9 PLC \\ & + \lambda_{10} INS + \lambda_{11} ASR + \lambda_{12} CAD + \lambda_{13} LAN \\ & + \lambda_{13+i} IND_i + \mu \end{aligned}$$

(10)

DATA AND ESTIMATION

Inclusion Criteria

According to the Malaysian Standard Industrial Classification 2000 (MSIC 2000) there are 53 three-digit industries. In order to obtain a rational outcome, the study needs to be conducted only within industries in which FMT is intensively adopted. On account of this, inclusion criteria were formulated in an effort to select FMT intensively adopted MSIC three-digit industries for the sample, which is shown below:

Industries with high "capital/labour" ratio.

Industries in which product variation is a marketing strategy.

Industries in which products are susceptible to demand fluctuation.

Using the above criteria a sample of 16 MSIC threedigit industries which together comprise 50 five-digit industries was selected.

Primary Data

The data that indicate the degree of adoption of FMT is not published by any organisation in Malaysia. Hence, a questionnaire survey was conducted to gather information necessary to compute the percentage of establishments adopting each specific type of FMT in a given year, within a given MSIC five-digit industry. The questionnaires were sent to all the establishments, listed under the 50 MSIC five-digit industries that appeared in the directory of Federation of Malaysian Manufacturers.

Secondary Data

In order to compute TE, industry-wise data is required for output, intermediate input, capital input and labour input. The closest indicators for these values were obtained from of the ASMI published for the years 2000 through 2005 by the DOS of Malaysia. The variables GLO was computed using the data obtained from this table. OWNFISZE and INC were computed using the data obtained from the Economic Censes conducted by the DOS Malaysia. The information required to calculate FAGE was obtained from the Directory published by the Federation of Malaysian Manufacturers.

EMPIRICAL RESULTS

Multicollinearity of FMT

Since only FMT intensively used industries were included in the sample, naturally some similarity in the sequence and characteristics of the production processes could be expected even amongst different five-digit industries. Hence, there could be a tendency for a similarity in the technology adopted amongst these industries. Due to the similarities in technologies, a high prevalence of multicollinearity among the eight types of FMT could be anticipated. In this study, bivariate Pearson productmoment correlation analysis has been conducted using SPSS to test for multicollinearity amongst FMT. The output that reveals potential multicollinearity among FMT variables is displayed in Table 1. According to Coakes, Steed and Price [29] and Field [30], when a considerable number of correlations are exceeding 0.3, the matrix is suitable for Principal Component Analysis (PCA).

PCA was performed using SPSS in order to obtain underlying dimensions (Principal Components) of FMT as a remedy for multicollinearity. As per both standard methods of (i.e. screen test and eigen values greater than one) extracting the optimal number of components, three Principal Components (PCs) were extracted that account for 67 percent of the variation in the FMT. According to Table 2, the loadings of variables onto the three PCs obtained from both types of rotations (Orthogonal and Oblique) are quite similar. Hence, due to simplicity, PCs obtained from orthogonal rotation was used in the rest of the analysis.

Once the most appropriate type of rotation and the resultant PCs were decided, the variables loading onto each of these PCs were examined as the next step. An examination of the component loadings depicted in Table 2 indicates that LAN, CAD, PLC and CNC load onto PC1; ASR, INS and ROB load onto PC2 while only NC loads onto PC3. Usually it is difficult to give clear cut themes or names to PCs that only relate to or encompass particular variables that are loading onto it. Hence, only the best possible names have been assigned to the PCs extracted from this analysis. The technologies LAN, CAD, PLC and CNC are used in the manufacturing set up as process control technologies. Since these load onto PC1,

Table 1. Correlations among FMT

		CNC	NC	ROB	PLC	INS	ASR	CAD	LAN
CNC	Pearson Correlation	1.000	0.160**	0.351**	0.634**	0.307**	0.237**	0.248**	0.322**
	Sig. (2-tailed)		0.005	0.000	0.000	0.000	0.000	0.000	0.000
	N	300	300	300	300	300	300	300	300
NC	Pearson Correlation	0.160**	1.000	0.012	0.164**	0.177**	0.126*	0.141*	0.171**
	Sig. (2-tailed)	0.005		0.836	0.005	0.002	0.030	0.014	0.003
ROB	Pearson Correlation	0.351**	0.012	1.000	0.368**	0.250**	0.427**	0.391**	0.236**
	Sig. (2-tailed)	0.000	0.836		0.000	0.000	0.000	0.000	0.000
PLC	Pearson Correlation	0.634**	0.164**	0.368**	1.000	0.302**	0.257**	0.394**	0.380**
	Sig. (2-tailed)	0.000	0.005	0.000		0.000	0.000	0.000	0.000
INS	Pearson Correlation	0.307**	0.177**	0.250**	0.302**	1.000	0.564**	0.115*	0.186**
	Sig. (2-tailed)	0.000	0.002	0.000	0.000		0.000	0.046	0.001
ASR	Pearson Correlation	0.237**	0.126*	0.427**	0.257**	0.564**	1.000	0.308**	0.129*
	Sig. (2-tailed)	0.000	0.030	0.000	0.000	0.000		0.000	0.025
CAD	Pearson Correlation	0.248**	0.141*	0.391**	0.394**	0.115*	0.308**	1.000	0.609**
	Sig. (2-tailed)	0.000	0.014	0.000	0.000	0.046	0.000		0.000
LAN	Pearson Correlation	0.322**	0.171**	0.236**	0.380**	0.186**	0.129*	0.609**	1.000
	Sig. (2-tailed)	0.000	0.003	0.000	0.000	0.001	0.025	0.000	

**Correlation is significant at the 0.01 level (2-tailed);*Correlation is significant at the 0.05 level (2-tailed).

Table 2. Comparison of components obtained from two types of rotations

	Component One			Component Two			Component Three		
	Orthogonal	Oblique		Orthogonal	Oblique		Orthogonal	Oblique	
		Pattern	Structure		Pattern	Structure		Pattern	Structure
LAN	0.816	0.861	0.811						
CAD	0.816	0.841	0.801						
PLC	0.666	0.640	0.722			0.445			
CNC	0.555	0.517	0.621			0.467			
ASR				0.845	0.858	0.851			
INS				0.816	0.844	0.826			
ROB	0.477	0.412	0.542	0.526	0.460	0.573			
NC							0.883	0.871	0.883

so can be named as "process control" technologies. The technologies ASR, INS and ROB load onto PC2, so can be named as "production and quality control" technologies. PC3 has only one variable i.e. NC, loading onto it so can be called the "general control" technology.

As the next step, the eight FMT variables were substituted with the three PCs namely, PC1, PC2 and PC3. Therefore, the TE model was reformulated as follows:

$$TE = \lambda_0 + \lambda_1 GLO + \lambda_2 OWN + \lambda_3 FSIZE$$
$$+ \lambda_4 FAGE + \lambda_5 INC + \lambda_6 PC1 + \lambda_7 PC2$$
$$+ \lambda_8 PC3 + \lambda_{8+i} IND_i + \mu$$

Multiple Regression Analysis of TE

As described, the model contains a set of 16 industry fixed dummy variables (IND_j) to account for the differences of technological opportunity among industries. Although it is theoretically desirable to include IND_j, the consequent impact of adding these 16 extra variables needs to be examined by comparing and contrasting the results obtained without considering the IND_j in the model. A separate regression was performed for this scenario and the tables of Model Summary, ANOVA and Coefficients were obtained. In order to facilitate the easy comparison of the results, the tables of output obtained from regression analysis for the two situations, one with the IND_j included and the other without the IND_j have been combined into one. The tables of Model Summary, ANOVA and Coefficients contained in the SPSS output for these situations have been reproduced in Tables 3-5 respectively.

According to **Table 3**, Adjusted R square is considerably high (0.983) when IND_j has been included. This indicates that the explanatory variables together explain 51.8 percent of the variance in TE. However, the explanatory power of the model has decreased significantly when the IND_j has been excluded; Adjusted R square (0.271) has decreased.

According to ANOVA, the F statistics for both situations of including and excluding IND_j in the models are 10.905 and 14.884 respectively. They both are larger than the critical value (1.569) of the F distribution, obtained from the F distribution calculator for $\alpha = 0.05$ level of significance when degrees of freedom are 23 and 276.

The statistical test for the existence of a linear relationship between dependent variable and the independent variables is:

$$H_0 : \lambda_1 = \lambda_2 = \lambda_3 = \cdots = \lambda_k = 0$$

H_1 : Not all $\lambda_i \left(i = 1, 2, \cdots, 24 \right)$ are zero

As the F statistic is in the rejection region, H_0 was rejected and H_1 was accepted. Since "p < 0.000", it can be concluded that there is strong evidence of TE having a linear regression relationship with any of the explanatory variables in the model with a probability of less than 0.1 percent of making an error in this conclusion.

IND_j Included

According to **Table 5**, the variables namely, GLO (0.000) and INC (0.000) are very highly significant at "p < 0.001". This implies that chances of making an error by assuming that these variables correlate with TE is less than 0.1 percent. Also both variables show a positive relationship with the dependent variable. FSIZE (0.136) is marginally significant at "$0.10 < p < 0.15$" and positively correlated whereas OWN (0.199) is insignificant and positively correlated. However, FAGE (0.889) is very highly insignificant. Since the main focus of this model is to test the

Table 3. Model summary[b]

R		R Square		Adjusted R Square		Std. Error of the Estimate	
IND_j Included	IND_j Excluded	IND_j Included	IND_j Excluded	IND_j Included	IND_j Excluded	IND_j Included	IND_j Excluded
0.645	0.418	0.416	0.175	0.367	0.152	0.062700	0.072548

[b]Dependent Variable: TE.

Table 4. ANOVA[b]

	Sum of Squares		df		Mean Square		F		Sig. (p-value)	
	Indj included	Indj excluded	Indj included	Indj excluded	Indj included	Indj excluded	Indj included	Indj excluded	Indj included	Indj excluded
Regression	8.784	3.337	23	8	0.382	0.070	10.905	14.884	0.000	0.000
Residual	9.666	13.093	276	291	0.035	0.045				
Total	18.450	18.450	299	299						

[b]Dependent Variable: TE.

Table 5. Coefficients[a]

Variable	IND$_j$ included		IND$_j$ excluded	
	B	Sig. (p-value)	B	Sig. (p-value)
(Constant)	0.526	0.000	0.567	0.000
GLO	0.000	0.000	0.000	0.000
OWN	0.054	0.199	−0.082	0.012
FSIZE	2.702E−11	0.136	−2.943E−11	0.035
FAGE	0.000	0.889	−0.002	0.565
INC	1.955	0.000	1.669	0.000
PC1	0.022	0.069	0.036	0.003
PC2	0.015	0.097	0.024	0.008
PC3	−0.038	0.005	−0.058	0.000

[a]Dependent Variable: TE.

significance of the correlation of FMT with TE, an examination of the correlation of the three PCs with TE becomes necessary. Both PC1 (0.069) and PC2 (0.097) are moderately significant at "0.1 < p < 0.05" and positively correlated with TE. Although PC3 (0.005) is highly significant, its relationship with TE is negative.

IND$_j$ Excluded

A separate regression was performed for the scenario in which 16 IND$_j$ variables were excluded from the TE model and the tables of Model Summary, ANOVA and Coefficients contained in the respective SPSS output have been reproduced in the respective tables for the scenario of IND$_j$ excluded. The Adjusted R square (0.271) has decreased considerably. According to the output of the model that excluded the IND$_j$, the significance of explanatory variables, GLO, OWN, FSIZE, FAGE, INCPC1, PC2 and PC3 are 0.000, 0.012, 0.035, 0.565, 0.000, 0.003, 0.008 and 0.000 respectively. One crucial difference in this model is contrary to a priori expectations, both OWN and FSIZE have negative coefficients. However, PC1 and PC2 are highly significant at "0.001 < p < 0.01" which are only moderately significant according to the model in which IND$_j$ variables were included. However, after considering all attributes it is inferred that the reliability of the first TE model which included IND$_j$ variable is higher.

For all the cases, Mahalanobis distance and Cooks distance which indicate the impact of outliers had been saved in the SPSS data editor. The critical chi-square value of 51.179 at $\alpha = 0.001$ level of significance was taken as the critical value for the Mahalanobis distance. There were 19 cases which exceeded the critical value indicating that there were 19 multivariate outliers among the 300 cases. The critical value considered for the Cooks distance was one and only in two cases the critical value was exceeded.

The variables, PC1 and PC2 of the FMT which represent two important themes (dimensions), namely "process control technologies" and "production and quality control technologies" which together account for 53 percent of the variance in FMT is significant. The third PC which represents "general control technology" is insignificant and it only accounts for 13 percent of the variance in FMT. According to both TE models, the null hypotheses that PC1 and PC2 have no partial correlation with TE (i.e. $\lambda_6 = 0$, and $\lambda_7 = 0$) can be rejected. Moreover, the TE model which included IND_j (the more reliable model), the relationship both PC1 and PC2 have with the TE is moderately significant. Therefore, the alternative hypotheses can be accepted which means that FMT has a significant correlation with TE which is positive (since in both models λ_6 and λ_7 are positive). This leads to the acceptance of the research hypothesis: A high degree of FMT adoption enhances TE of the Manufacturing Industry of Malaysia.

CONCLUSIONS

The aim of this paper was to examine the impact of the degree of adoption of FMT on the technical efficiency of the manufacturing industry of Malaysia. The types of FMT considered were namely, Computer Numerical Control machine tools (CNC), Numerical Controlled Machine Tools (NC), Robotics (ROB), Programmable Logic Controllers (PLC), Automated Inspections (INS), Automated Storage and Retrieval Systems (ASR), Computer Aided Design (CAD) and Local Area Networks (LAN). In order to remove the effects of multicollinearity among the eight types of FMT they were substituted in the regression model with the three PCs. The FMT variables load onto PCs as follows: LAN, CAD, PLC and CNC load onto PC1; ASR, INS and ROB load onto PC2 and NC only loads onto PC3. The three PCs were labelled so that they best describe their respective constituents; PC1—"process control" technologies, PC2— "production and quality control" technologies and the PC3—"general control" technology.

The most important finding of the study is that both PC1 and PC2 show moderately significant and positive relationships with TE. This indicates that the increasing adoption of process control technologies and production and quality control technologies have direct impact on TE of the FMT intensively adopted sub sector of the manufacturing industry. In contrast, PC3 shows a very highly significant and negative relationship with TE. Since both PC1 and PC2 together account for greater variation (53 percent) and PC3 account for relatively smaller variation (12 percent) among the eight FMT, it can be concluded that a high degree of FMT adoption enhances TE of the manufacturing industry of Malaysia. This is consistent with the a priori expectations regarding FMT.

This shows that the degree of adoption of FMT has a positive relationship with embodied technological change which captures the effects of learning by doing (experience), advances in applied technology, managerial efficiency and industrial organisation which affords better methods and organisations that improve the efficiency of both new and old factor inputs. In other words, empirical findings of the present study postulate a correlation between the degree of FMT adoption and the above stated effects. Although the empirical findings suggest prevalence of a greater efficiency within the FMT intensively adopted industries, it does not indicate which specific effect is causing the efficiency. In the light of the deductions made in this section it can be stated that the higher the degree of adoption of FMT, the greater will be the ability to manufacture more output with the same factor inputs, in other words the ability to produce cost effectively.

In its usual call for future research, the authors recommend studies that investigate the relationship of investments in FMT rather than the degree of FMT adoption have with the TE of the manufacturing industry. It can be safely admitted that the accuracy of findings can be increased considerably by considering investments in FMT rather than the degree of adoption of FMT. Hence, it is proposed that future studies need be undertaken in collaboration with the statuary bodies established to oversee and facilitate the manufacturing industry which makes establishments obligatory to divulge investments made in FMT in order to evaluate the impact of investments in FMT on TE.

REFERENCES

1. R. Mahadevan, "Perspiration versus Inspiration: Lessons from a Rapidly Developing Economy," Journal of Asian Economics, Vol. 18, No. 2, 2007, pp. 331-347.doi:10.1016/j.asieco.2007.02.009

2. T. J. Coelli, D. S. P. Rao, C. J. O'Donnell and G. E. Battese, "An Introduction to Efficiency and Productivity Analysis," Springer, New York, 2005.

3. D. J. Aigner, C. A. K. Lovell and P. J. Schmidt, "Formulation and Estimation of Stochastic Frontier Production Models," Journal of Econometrics, Vol. 6, No. 1, 1977, pp. 21-37.doi:10.1016/0304-4076 77 90052-5

4. R. Mahadevan, "A Frontier Approach to Measuring Total Factor Productivity Growth in Singapore's Services Sector," Journal of Economic Studies, Vol. 29, No. 1, 2002, pp. 48-58. doi:10.1108/01443580210414111

5. R. Mahadevan and K. P. Kalirajan, "On Measuring Total Factor Productivity Growth in Singapore's Manufacturing Industries," Applied Economics

Letters, Vol. 6, No. 5, 1999, pp. 295-298. doi:10.1080/135048599353267

6.   J. Wind and A. Rangaswamy, "Customisation: The Next Revolution in Mass Customization," Journal of Interactive Marketing, Vol. 15, No. 1, 2001, pp. 13-31.

7.   G. Da Silveria and F. S. Fogliatto, "Effects of Technology Adoption on Mass Customisation Ability of Broad and Narrow Market Firms," Gestao & Producao, Vol. 12 No. 3, 2005, pp. 347-359. doi:10.1590/S0104-530X2005000300006

8.   P. Kumar and S. G. Deshmukh, "A Model for Flexible Supply Chain through Flexible Manufacturing," Global Journal of Flexible Systems Management, Vol. 7, No. 3-4, 2006, pp. 17-24.

9.   R. Sinha and C. H. Noble, "The Adoption of Radical Manufacturing Technologies and Firm Survival," Strategic Management Journal, Vol. 29, No. 9, 2008, pp. 943- 962.doi:10.1002/smj.687

10.   Malaysian Industrial Development Authority (MIDA), "Malaysia-Investment in the Manufacturing Sector," MIDA, 2007.

11.   C. Sun and K. L. Kalirajan, "Gauging the Sources of Growth of High-Tech and Low-Tech Industries: The Case of Korean Manufacturing," Blackwell Publishing Ltd., Hoboken, 2005.

12.   X. Zhang and S. Zhang, "Technical Efficiency in China's Iron and Steel Industry: Evidence from the New Census Data," International Review of Applied Economics, Vol. 15, No. 2, 2001, pp. 199-211. doi:10.1080/02692170151137078

13.   J. D. Lee, T. A. Kim and E. Heo, "Technological Progress versus Efficiency Gain in Manufacturing Sectors," Review of Development Economics, Vol. 2, No. 3, 1998, pp. 268-281. doi:10.1111/1467-9361.00041

14.   G. Battese and S. S. Broca, "Functional Forms of Stochastic Frontier-Production Functions and Models for Technical Inefficiency Effects: A Comparative Study for Wheat Farmers in Pakistan," Journal of Productivity Analysis, Vol. 8, No. 4, 1997, pp. 395-414.

15.   E. R. Brendt and C. J. Morrsison, "High Tech Capital Formation and Economic Performance in USA Manufacturing Industries: An Exploratory Analysis," Journal of Econometrics, Vol. 65, No. 1, 1995, pp. 9-43.

16.   D. A. R. Dolage, A. B. Sade and M. A. Elsadig, "The Influence of Flexible Manufacturing Technology Adoption on Productivity of Malaysian Manufacturing Industry," Economic Modelling, Vol. 27, No. 1, 2010, pp. 395- 403. doi:10.1016/j.econmod.2009.10.005

17.   L. H. Amato and C. H. Amato, "The Impact of High Tech Production

Techniques on Productivity and Profitability in Selected US Manufacturing Industries," Review of Industrial Organization, Vol. 16, No. 4, 2000, pp. 327-342.doi:10.1023/A:1007800121100

18. D. A. R. Dolage and A. B. Sade, "The Impact of Adoption of Flexible Manufacturing Technology on Price Cost Margin of Malaysian Manufacturing," Technology and Investment, Vol. 3, No. 1, 2012, pp. 26-35. doi:10.4236/ti.2012.31005

19. R. Mahadevan, "Is There a Real TFP Growth Measure for Malaysia's Manufacturing Industries?" ASEAN Economic Bulletin, Vol. 19, No 2, 2002, pp. 178-190.doi:10.1355/AE19-2E

20. K. P. Kalirajan and R. T. Shand, "Frontier Production Functions and Technical Efficiency Measures," Journal of Economic Surveys, Vol. 13, No. 2, 1999, pp. 149-172.doi:10.1111/1467-6419.00080

21. R. Mahadevan, "Assessing the Output and productivity Growth of Malaysia's Manufacturing Sector," Journal of Asian Economics, Vol. 12, No. 4, 2001, pp. 587-597.doi:10.1016/S1049-0078 01 00104-X

22. Q. N. Vu, "Technical Efficiency of Industrial StateOwned Enterprises in Vietnam," Asian Economic Journal, Vol. 17, No. 1, 2003, pp. 87-101.

23. H. Margono and S. C. Sharma, "Efficiency and Productivity Analysis of Indonesian Manufacturing Industries," Journal of Asian Economics, Vol. 17, No. 6, 2006, pp. 979-995.doi:10.1016/j.asieco.2006.09.004

24. Federation of Malaysian Manufacturers (FMM), "Directory -2007," FMM, 2007.

25. R. Mahadevan, "How Technically Efficient Are Singapore's Manufacturing Industries?" Applied Economics, Vol. 32, No. 15, 2000, pp. 2007-2014.doi:10.1080/00036840050155931

26. S. Chandrasiri, "Productivity and Technology in Sri Lankan Manufacturing Industry, Human Development in a Knowledge-Based Society: Sri Lankan Scene," The Sri Lanka Economic Association, Colombo, 2005.

27. R. H. McGuckin and M. L. Streitwieser, "The Effect of Technology Use on Productivity Growth," Centre of Economic Studies, 1996.

28. D. K. Round, "Price Cost Margins in Australian Manufacturing Industries, 1971-1972," University of Adelaide, Adelaide, 2001.

29. S. J. Coakes, L. Steed and J. Price, "SPSS 15.0, Analysis without Anguish," John Wiley & Sons Australia, Ltd., Melbourne, 2008.

30. A. Field, "Discovering Statistics Using SPSS," SAGE Publications Ltd., London, 2005.

Chapter 2

NANOMATERIAL PRODUCTION AND DOWNSTREAM HANDLING PROCESSES

INTRODUCTION

The number of commercial applications of nanomaterials is growing at a tremendous rate. As this rapid growth continues, it is essential that producers and users of nanomaterials ensure a safe and healthy work environment for employees who may be exposed to these materials. Unfortunately, because nanotechnology is so new, we do not know or fully understand how occupational exposures to these agents may affect the health and safety of workers or even what levels of exposure may be acceptable. Given our current knowledge in this field, it is important to take precautions to minimize exposures and protect safety and health. This document discusses approaches and strategies to protect workers from potentially harmful exposures during nanomaterial manufacturing, use, and handling processes. Its purpose is to provide the best available current knowledge of how workers may be exposed and provide guidance on exposure control and evaluation. It is intended to be used as a reference by plant managers and owners who are responsible for making decisions regarding capital allocations, as well as health and safety professionals, engineers, and industrial hygienists who are specifically charged with protecting worker health in this new and growing field. Because little has been published on exposure controls in the production and use of nanomaterials, this document focuses on applications that have relevance to the field of nanotechnology and on engineering control technologies currently used, and known to be effective, in other industries. This document also addresses other approaches to worker protection, such as the use of administrative controls and personal protective equipment.

Background

Nanotechnology is the manipulation of matter at the atomic scale to create materials, devices, or systems with new properties and/or functions. Around the world, the introduction of nanotechnology promises great societal benefits across many economic sectors: energy, healthcare, industry, communications, agriculture, consumer products, and others [Sellers et al. 2009]. Some nanoparticles are natural, as in sea salt or pine tree pollen, or are incidentally produced, as in volcanic explosions or diesel engine emissions. The focus of this document is engineered nanomaterials, those materials deliberately engineered and manufactured to have certain properties and have at least one primary dimension of less than 100 nanometers (nm). Nanomaterials have properties different from those of their bulk components. For example, many of these materials have increased strength/weight ratios, enhanced conductivities, and improved optical or magnetic properties. These new properties make nanomaterial development so exciting and are the reason they hold the promise of great economic potential. Nanomaterials are often classified by their physicochemical characteristics or structure. The four classes of materials of which nanoparticles are typically composed include elemental carbon, carbon compounds, metals or metal oxides, and ceramics. The nanometer form of metals, such as gold, and metal oxides, such as titanium dioxide, are the most common engineered nanomaterials being produced and used [Sellers et al. 2009]. Nano-sized silica, silver, and natural clays are also common materials in use. The carbon nanotube is a unique nanomaterial being investigated for a wide range of applications. These tubes are cylinders constructed of rolled-up graphene sheets. Another interesting carbon structure is a fullerene (also known as a Bucky Ball). These are spherical particles usually constructed from 60 carbon atoms arranged as 20 hexagons and 12 pentagons. As shown in Figure 1, the structure resembles a geodesic dome (designed by architect Buckminster Fuller, hence the name). Nanomaterials are widely used across industries and products, and they may be present in many forms. Significant international health and safety research and guidance concerning the handling of nanomaterials is underway to support risk management of commercial developments. Both risks and rewards are inherent in these new materials. Scientists around the world are conducting toxicological studies on these nanomaterials, and initial findings are concerning. Animals exposed to titanium dioxide (TiO_2) and carbon nanotubes (CNTs) have displayed pulmonary inflammation [Chou et al. 2008; Rossi et al. 2010; Shvedova et al. 2005]. Other studies have shown that nanoparticles can translocate to the circulatory system and to the brain and cause oxidative stress [Elder et al. 2006; Wang et al. 2008]. Perhaps the most troubling finding is that CNTs can cause asbestos-like pathology in mice [Poland et al. 2008; Takagi et al. 2008].

Figure 1: Atomic structure of a spherical fullerene.

Industry Overview

In March 2006, the Woodrow Wilson International Center for Scholars created an inventory of 212 consumer products or product lines that incorporate nanomaterials (http://www. nanotechproject.org/inventories /consumer/ analysis_draft/). These products were broken down into eight categories using a publically available consumer product classification system. As of March 2011, the number of consumer products has increased by 521% (212 to 1,317 nano-enabled products) with products coming from more than 24 nations [WWICS 2011]. These products include acne lotions, antimicrobial treatment for socks, sunscreens, food supplements, components for computer hardware (such as processors and video cards), appliance components, coatings, and hockey sticks. Of the current 1,317 nano-enabled products, the largest product category with 738 products was health and fitness. The most common type of nanomaterial used in these products was silver (313 products), followed by carbon (91 products) and titanium dioxide (59 products). Roco [2005] reports that worldwide, the investment in nanotechnology has increased from $432 million in 1997 to about $4.1 billion in 2005. In this same time period, the U.S. government investment in nanotechnology has increased to nearly $1.1

billion. Estimates made in 2000 suggested that $1 trillion in products will use nanotechnology in some way by 2015. The National Science Foundation estimates that the number of workers in this industry will increase to 2 million worldwide by 2015. Currently, most production facilities are relatively small, with lab, bench, or, at most, pilot plant operations [Genaidy et al. 2009]. This is also indicative of downstream users (applications and product development). As new manufacturing processes and technologies are developed and introduced, novel materials with unknown toxicological properties will require effective risk management approaches. As more of these products enter the market, concern about the health and safety of the workers grows.

Occupational Safety and Health Management Systems

Control measures for nanoparticles, dusts, and other hazards should be implemented within the context of a comprehensive occupational safety and health management system [ANSI/AIHA 2012]. The critical elements of an effective occupational safety and health management system include management commitment and employee involvement, worksite analysis, hazard prevention and control, and sufficient training for employees, supervisors, and managers (www.osha.gov/Publications/safety-health-management-systems.pdf). In developing measures to control occupational exposure to nanomaterials, it is important to remember that processing and manufacturing involve a wide range of hazards. Conducting a preliminary hazard assessment (PHA) encompasses a qualitative life cycle analysis of an entire operation, appropriate to the stage of development:

- Chemicals/materials being used in the process
- Production methods used during each stage of production
- Process equipment and engineering controls employed
- Worker's approach to performing job duties
- Exposure potential to the nanomaterials from the task/operations
- The facility that houses the operation

The steps taken to perform PHAs for specific operations should be documented to let others know what was done and to help others understand what works. PHAs are frequently conducted as initial risk assessments to determine whether more sophisticated analytical methods are needed and to prepare an inventory of hazards and control measures needed for these hazards. One or two individuals with a health and safety background and knowledge of the process can perform PHAs. As part of the assessment, the health and safety professional should evaluate the magnitude of the emissions (or potential emissions) and the effects of exposure to these emissions. PHAs are an important first step toward developing control measures that can be considered during the planning stage. Essentially, hazard control should be an integral component of facility, process, and equipment design and construction. This includes design for inherent process safety. The use of engineering controls should be considered as part of a comprehensive control strategy for hazards associated with processes/ tasks that cannot be effectively eliminated, substituted for, or contained through process equipment modifications. The standards for an occupational health and safety management system, as outlined in ANSI/AIHA Z10 [ANSI/AIHA 2012] and BSI 18001 [BSI 2007c], promote a continuous improvement cycle (plan, do, check, act), which does not have an exit point and is the basis for worksite analysis. Figure 2 illustrates how control measures are incorporated into an occupational safety and health management system. The continuous improvement loop is applicable to all hazards in a process/facility (e.g., airborne contaminant exposures, ergonomic, combustible dusts, fire safety, and physical hazards). The hazard assessment should be reviewed during each cycle described by Figure 2 and periodically updated when major changes occur. Although the optimal time to undertake a PHA is during the design stage, hazard assessments can also be done during the operation of a facility and have the benefit of using existing data. After the PHA is complete, the nanomaterial risk management plan is designed to avoid or minimize hazards discovered during the assessment. The following options should be considered:

Automated product transfer between operations. A process that allows for continuous process flow to avoid exposures caused by workers handling powdered or vaporous materials.

- Closed-system handling of powdered or vaporous materials, such as screw feeding or pneumatic conveying.

- Local exhaust ventilation. Steps should be taken to avoid having positive pressure ducts in work spaces because leakage from ducts can cause exposures. Ducts or pipes should be connected using flanges with gaskets that prevent leakage.

- Continuous bagging for the intermediate output from various processes and for final products. A process discharges material into a continuous bag that is sealed to eliminate dust exposures caused by powder handling. Bags are heat sealed after loading.

- Minimizing the container size for manual material handling. Minimizing the size of the container or using a long-handled tool is recommended so that the worker does not place his breathing zone inside the container (as shown in Figure 3). NIOSH recommends a maximum container depth of 25 inches [NIOSH 1997]. If large containers are required, engineering controls to provide a barrier between the container and the breathing zone of the worker are recommended.

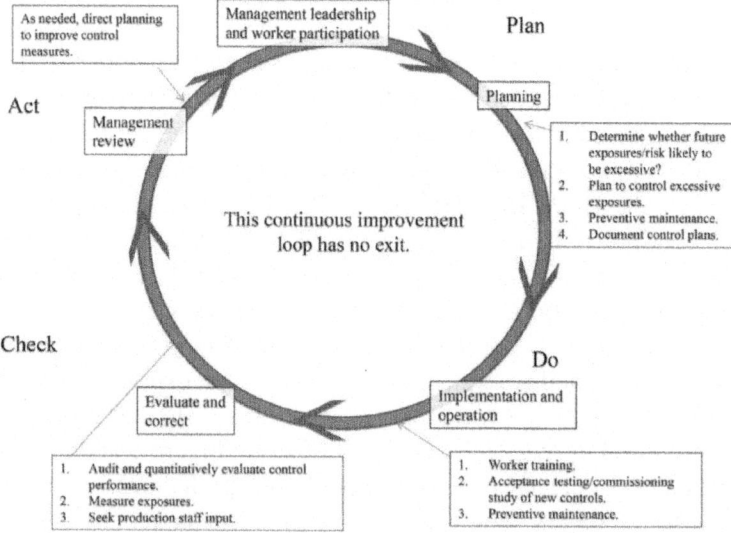

Figure 2: How control measures are selected, implemented, and managed into an occupational safety and health management system. (adopted from [ANSI 2005]).

Photo by NIOSH

Figure 3: Worker reaching into drum.

Many good resources are available on the occupational safety and health risk management of nanomaterials. Comprehensive documents have been produced by a number of organizations. Some of these are listed in Appendix A.

Prevention through Design (PtD)

The concept of Prevention through Design (PtD) is to design out or minimize hazards, preferably, early in the design process. PtD is also called inherent or intrinsic safety, safety by design, design for safety, and safe design. When PtD is implemented, the control hierarchy is applied by designers (e.g., engineers, architects, industrial designers) and business leaders (e.g., owners, purchasers, managers) who consider the benefits of designing safety into things external to the worker to prevent work-related injuries and illnesses. PtD strategies, like the hierarchy of controls, can take many forms. Elimination and substitution measures are desirable, but these strategies may be difficult to implement when working with nanomaterials because these materials are likely being used for their unique properties. The pharmaceutical industry has addressed some of these challenges since their products must be contained rather than removed or eliminated from the process. They have adopted a containment hierarchy of controls that addresses designing inherent safety into the process [Brock 2009]. The initial levels of containment include elimination and substitution as well as product, process, and equipment modifications. Only after efforts have been made to design the process to reduce potential emissions sources should engineering controls be considered. Other PtD strategies can be considered:

- Limiting process inventories by producing the nanomaterials as they are consumed in the process.
- Operating a process at a lower energy state (e.g., lower temperature or pressure), which typically results in lower fugitive emissions and therefore safer operation.
- Using fail-safe devices where possible. Fail-safe devices are designed so that if they fail, the system reverts to a safer condition. An example of a fail-safe device is a valve controlling a reagent for a reaction. If the safe condition for the system is for this valve to be closed, the fail-safe valve would automatically close in the event of a failure.
- Installing a closed transport system to eliminate worker exposures during transport activities

PtD strategies typically do not include administrative controls and personal protective equipment (PPE) as the primary controls. These measures require worker interaction with the process or active steps to limit the extent of the hazard. Most effective PtD approaches reduce or eliminate hazardous conditions without relying on input from workers. Humans are generally recognized as being much less reliable than most machines, particularly in emergencies [Kletz 2001]. The use of administrative controls and PPE in PtD strategies is generally for redundancy—further safeguards should the primary control fail.

The ideal time to develop a PtD strategy is during the development phase of a process, material, or facility. As the nanotechnology field is still in its relative infancy, there are numerous opportunities to implement PtD in the early stages. The manner in which these materials are handled and processed can largely affect the overall safety of the process, and the health and safety of workers may be significantly improved through the implementation of a PtD strategy

OELs as Applied to Nanotechnology

Occupational exposure limits (OELs) are useful in reducing work-related health risks by providing a quantitative guideline and basis to assess the worker exposure potential and the performance of engineering controls and other risk management approaches. Currently, no regulatory standards for nanomaterials have been established in the United States. However, NIOSH has recently published two current intelligence bulletins (CIBs) regarding occupational exposures to nanomaterials. In a CIB on titanium dioxide (TiO_2), NIOSH recommends exposure limits of 2.4 mg/m3 for fine TiO_2 and 0.3 mg/m3 for ultrafine (including engineered nanoscale) TiO_2 , as time-weighted average (TWA) concentrations for up to 10 hours per day during a 40-hour work week [NIOSH 2011]. In a CIB on carbon nanotubes and nanofibers, NIOSH recommends that worker exposure be limited to no more than 1 µg/m3 [NIOSH 2013]. Other countries have established OELs for various nanomaterials. For example, the British Standards Institute recommends working exposure limits for nanomaterials based on various classifications such as solubility, shape, and potential health concerns as related to larger particles of the same substance [BSI 2007b]. Germany's Institut für Arbeitsschutz der Deutschen Gesetzlichen Unfallversicherung, an institute for worker safety, has published similar guidelines [IFA 2009]. In the absence of governmental or consensus guidance on exposure limits, some manufacturers have developed suggested OELs for their products. For example, Bayer has established an OEL of 0.05 mg/m3 for Baytubes® (multiwalled CNTs) [Bayer MaterialScience 2010]. For Nanocyl CNTs, the no-effect concentration in air was estimated to be 2.5 µg/m³ for an 8-hr/day exposure [Nanocyl 2009]. Another approach that may be taken when OELs are absent is the ALARA concept, As Low As Reasonably Achievable. While ALARA is generally the goal for all occupational exposures, this concept is particularly useful when OELs are absent or in the case of contaminants with unknown toxicity

Control Banding

Control banding is a qualitative risk characterization and management strategy, intended to protect the safety and health of workers in the absence of chemical and workplace standards. Control banding groups workplace risks into hazard bands based on evaluations of hazard and exposure information [NIOSH 2009b]. Note that control banding is not intended to be a substitute for OELs and does not alleviate the need for environmental monitoring or industrial hygiene expertise.

To determine the appropriate control scheme, one should consider the characteristics of the substance, the potential for exposure, and the hazard associated with the substance. Four main control bands, based on an overall risk level, have been developed:

- Good industrial hygiene (IH) practice, general ventilation, and good work practices
- Engineering controls including fume hoods or local exhaust ventilation
- Containment or process enclosure allowing for limited breaks in containment
- Special circumstances requiring expert advice

One basic principle of control banding is the need for a method that will return consistent, accurate results even when performed by nonexperts. Other requirements include having a user friendly strategy, readily available required information (e.g., material safety data sheet [MSDS]), practical guidance on applying the strategy, and worker confidence in the results. With the absence of OELs, control banding can be a useful approach in the risk management of nanomaterials [Maynard 2007; Schulte et al. 2008; Thomas et al. 2006; Warheit et al. 2007]. Several control banding tools are available for use with engineered nanomaterials. The CB Nanotool, for example, bases the control band for a particular task on the overall risk level, which is determined by a matrix that uses severity scores and probability scores [Paik et al. 2008]. The severity score is based on the toxicological effects of the nanomaterial, while the probability score relates to the potential for employee exposure. The health hazard categories for some control banding approaches are based upon the European Union risk phrases, while exposure potentials include the volume of the chemical used and the likelihood of airborne materials, estimated by the dustiness or volatility of the source compound [Maidment 1998]

EXPOSURE CONTROL STRATEGIES AND THE HIERARCHY OF CONTROLS

Controlling exposures to occupational hazards is the fundamental method of protecting workers. Traditionally, a hierarchy of controls has been used as a means of determining how to implement feasible and effective controls. Figure 4 shows one representation of this hierarchy. The idea behind the hierarchy of controls is that the control methods at the top of the triangle are generally more effective in reducing the risk associated with a hazard than those at the bottom. Following the hierarchy normally leads to the implementation of inherently safer systems, ones where the risk of illness or injury has been substantially reduced. Designing out hazards early in the design process is a basic tenet of PtD. When PtD is implemented, the control hierarchy is applied by designers and owners/managers to include safety into the process. The following sections discuss each element of the hierarchy of controls—elimination, substitution, engineering controls, administrative controls, and PPE— and how it may relate to nanotechnology.

Elimination

Elimination and substitution are generally most cost effective if implemented when a process is in the design or development stage. If done early enough, implementation is simple and, in the long run, can result in substantial savings (e.g., cost of protective equipment, first cost and operational cost for ventilation system). For an existing process, elimination or substitution may require major changes in equipment and/or procedures in order to reduce a hazard.

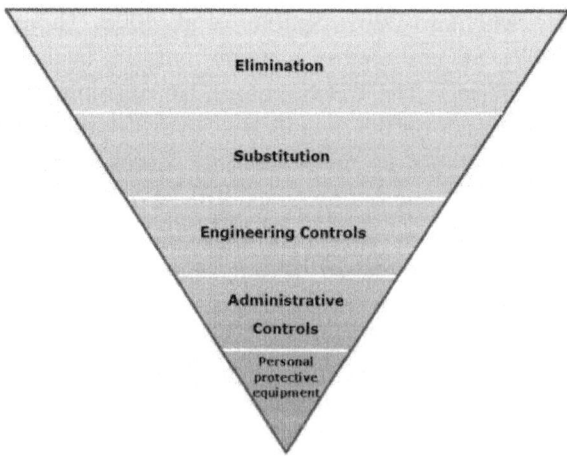

Figure 4: Graphical representation of the hierarchy of controls

Elimination is the most desirable approach in the hierarchy of controls. As its name implies, the idea behind elimination is to remove the hazard. Eliminating hazards is generally easiest to accomplish at the design stage, while the material, process, and/or facility is being developed. An example of elimination in a process step might be the removal of an incoming inspection step for nanomaterials. An incoming inspection that requires opening a package containing nanomaterials leads to the potential of aerosolization of those materials and therefore a potential hazard to the inspector. Eliminating the inspection step removes the hazard, thus creating an inherently safer process.

Substitution

Within the hierarchy of controls, the purpose of substitution is to replace one set of conditions having a high hazard level with a different set of conditions having a lower hazard level. Examples of substitution could include replacing a solvent-based (i.e., flammable) material with a water-based material, substituting a highly toxic material for one of lower toxicity, or changing a process's operating conditions so they are less severe (e.g., reduced pressure). Substitution of a nanomaterial may be difficult since it was likely introduced for its particular properties; however, some substitution may be possible. Substituting a nanomaterial slurry for a dry powder version will reduce aerosolization and provide a level of protection for workers handling the material. The specific nanomaterial should also be assessed because in some cases a less hazardous nanomaterial may provide the desired performance.

Engineering Controls

Engineering controls protect workers by removing hazardous conditions (e.g., local exhaust ventilation that captures and removes airborne emissions) or placing a barrier between the worker and the hazard (e.g., isolators and machine guards). Well-designed engineering controls can be highly effective in protecting workers and will typically be passive, that is, independent of worker interactions. It is important to design engineering controls that do not interfere with the productivity and ease of processing for the worker. If engineering controls make the operation more difficult, there will be a strong motivation by the operator to circumvent these controls. Ideally, engineering controls should make the operation easier to perform rather than more difficult. A good mantra in designing engineering controls is to "make it easier to do it the safe way." This also applies to administrative controls that are discussed later. The initial cost of engineering controls can be higher than administrative controls or personal protective equipment (PPE); however, over the long term, operating costs are frequently lower and, in some instances, can provide a cost savings in other areas of the process. The major benefit of engineering controls over administrative controls or PPE is, however, the inherent safety of the worker under a variety of conditions and stress levels. The use of engineering controls reduces the potential for worker behavior to impact exposure levels. Thus, when elimination and substitution are not viable options, the most desirable alternative for mitigating occupational hazards is to employ engineering controls. Engineering controls are likely the most effective and applicable control strategy for most nanomaterial processes.

In most cases, they should be more feasible than elimination or substitution and, given the potential toxicity of many nanomaterials, should prove to be more protective than administrative controls and PPE. Engineering controls are divided into two broad categories for discussion below: ventilation and nonventilation controls.

Ventilation The general concept behind ventilation for controlling occupational exposures to air contaminants, including nanomaterials, is to remove contaminated air from the work environment. The efficiency of the ventilation system can be affected by its configuration and flow volumes of both the air supplied to and the air exhausted from the work space. Effective ventilation applies to a wide range of applications including office heating, ventilating, and air conditioning (HVAC); infection control in healthcare; and control of emissions in industrial processes. Ventilation for occupant comfort, HVAC, is a specialized application of dilution ventilation and is not within the scope of this document. Filtration is a topic directly affecting ventilation; exhaust air laden with nanomaterials may need to be cleaned before being released into the environment. General ventilation can be used to achieve several goals for workplace contaminant control. A properly designed supply air ventilation system can provide plant ventilation, building pressurization, and exhaust air replacement. As new local exhaust hoods are installed in the production area, it is important to consider the need for replacement air, the location of the hood installation, and the need to rebalance the ventilation system. In general, it is necessary to balance the amount of exhausted air with a nearly equal amount of supply air. Without this replacement air, uncontrolled drafts will occur at doors, windows, and other openings; doors will become difficult to open due to the high pressure difference, and exhaust fan performance may degrade. In addition, turbulence created through high pressure differentials can defeat the design intent of the ventilation. Placement of the air supply registers in relation to other exhaust ventilation systems is important so that they do not negatively impact the desired performance. The use of general ventilation for dilution of contaminants being generated in the space should be restricted in its use depending on several factors discussed below. General ventilation used for dilution of contaminants by its nature is inefficient. One of two methods, recirculated air or single-pass air, may be used for this purpose. As the terms imply, recirculated air involves the treatment of exhaust air prior to its being returned to the area from which it was exhausted. Single-pass air is exhausted to the outside and may or may not require treatment prior to discharge. Both of these methods are expensive—the treatment of the recirculated air involves both first-cost and operating-cost penalties, while makeup-air treatment for single-pass air is inherently costly. According to the American Conference of Governmental Industrial Hygienists (ACGIH) Industrial Ventilation: A Manual of Recommended Practice for Design (hereafter referred to as the Industrial Ventilation Manual), dilution ventilation (i.e., air changes) to control exposure should be used only under specific conditions. Dilution ventilation for controlling health hazards is restricted by four limiting factors: (1) the quantity of contaminant generated must

not be too great or the airflow rate necessary for dilution will be impractical, (2) workers must be far enough away from the contaminant source or the evolution of contaminant must be in sufficiently low concentrations so that workers will not have an exposure in excess of the established threshold limit values (TLV®), (3) the toxicity of the contaminant must be low, and (4) the evolution of contaminants must be reasonably uniform [ACGIH 2013]. There are several issues with using dilution ventilation to control nanomaterial concentrations, including (1) there are no occupational exposure limits (TLVs mentioned above) or health effects data for many of the nanomaterials, (2) the toxicology data from some nanomaterials indicate that they may be associated with adverse health effects, and (3) it is difficult or impossible to calculate proper air change rates for contaminant control due to the variability in most operations. Therefore, local exhaust ventilation and good work practices should be used for controlling exposure, and air change rates should be based on the heat load requirements, general air movement, and comfort needs. The use of supply air for maintaining proper pressurization between production and nonproduction areas is a reasonable approach to reducing the exposure to nanomaterials outside of the immediate work zone. The fugitive emissions from nanomaterial production and processing may result in high background concentrations in the production area. When adjacent plant areas are nonproduction areas (e.g., office, quality assurance/control labs) or production areas where nanomaterials are not used, infiltration of nanoparticles may occur and result in the exposure of workers in those areas. Therefore, a negative air pressure differential should be maintained in the nanomaterial production area with respect to adjacent rooms/areas. This will help reduce the potential migration of airborne nanomaterials and exposure to other workers in adjacent rooms or areas. To maintain a slight negative pressure, the room supply air volume should be slightly less than the exhaust air. A general guide is to set a 5% flow difference between supply and exhaust flow rates but no less than 50 cfm [ACGIH 2013]. As with any good engineering control, a real-time monitor of differential pressure between areas should be employed, preferably with the control capability to modify airflows to maintain the required pressure differential

Local Exhaust Ventilation Local exhaust ventilation (LEV) is the application of an exhaust system at or near the source of contamination. If properly designed, it will be much more efficient at removing contaminants than dilution ventilation, requiring lower exhaust volumes, less make-up air, and, in many cases, lower costs. By applying exhaust at the source, contaminants are removed before they get into the general work environment. When designing a local exhaust ventilation system, it is important to understand the transport mechanisms of the contaminants that are to be removed. This will allow the design to use optimal flow rates and capture locations, maximizing the contaminant capture while minimizing impact on the process and reducing operating costs. LEV typically involves five components [Washington State L & I, no date]

- Exhaust hood. Examples include an enclosing hood to contain the contaminant, a receiving hood to capture or receive a contaminant that is released at a high velocity (e.g., grinding swarf), or simply an open duct.

- Duct. Transports the contaminant through the exhaust ventilation system.

- Air cleaner. Reduces the concentration of the contaminant in the exhaust air stream; may or may not be required.

- Fan. Moves the air through the exhaust system.

- Exhaust stack. Installed where the exhaust system discharges the air.

The exhaust hood captures the contaminant released by the process. It should be designed for the specific process being controlled, an important consideration for hot processes and those processes generating contaminants at high velocities. In either case, induced air flow (from high velocity air streams or rising air from a hot process) can overwhelm an insufficiently designed hood and allow contaminants to escape into the work environment. An important hood design factor is the capture velocity. This is the velocity of air needed to overcome contaminant velocity as well as room air currents. ACGIH Industrial Ventilation Manual contains a large collection of industrial ventilation hood designs for a wide selection of industrial processes [ACGIH 2013]. Though many of these designs have not been tested with nanomaterials, most are expected to perform effectively with these materials. An important consideration in hood design with nanomaterials is to provide the appropriate flow rates to prevent fugitive emissions without causing conditions that will remove nanomaterials from the process stream. Because of their very low mass, entrainment of nanomaterials in airflow streams occurs much more readily than with higher-mass particles.

Duct systems transport air between the various components of the LEV system. Designing duct systems requires balancing several factors. Duct losses caused by friction will increase with higher duct velocities, resulting in increased fan requirements and higher energy consumption; however, using larger ducts (in an effort to reduce duct velocity) results in increased duct purchase costs. A detailed method for designing and sizing LEV duct systems is provided by ACGIH [ACGIH 2013]. The choice of duct materials and sealing methods is particularly important when dealing with nanomaterials. The duct material needs to be impervious to the nanomaterials and suitable for use with nanomaterials having increased reactivity. The joints in the ducts should be sealed in such a way as to contain the nanomaterials.

Fans move air throughout the LEV system. Fans need to be sized to ensure adequate air flow while overcoming the system pressure drop (i.e., resistance to flow). Pressure drop is encountered when air is accelerated, such as within a hood; through ductwork due to frictional losses, particularly in fittings such as elbows; and through filters and other air-cleaning devices. Fan selection affects not only the control effectiveness of the LEV system but also its energy consumption. The fan system and the make-up air conditioning are typically the two greatest energy-consuming components of an LEV system. Proper fan selection needs to balance both control performance and operating efficiency [ACGIH 2013]. The same leakage and reactivity factors mentioned in the section on ductwork apply to fan selection.

Air cleaning is an important component of the LEV system, particularly if the exhaust air is returned to the building environment. Air cleaning involves the removal of gases and vapors, often with scrubbers and sorbent systems; however, in the case of nanomaterials, particulate removal systems will be required to eliminate them from the air stream. Cyclones, scrubbers, and other similar systems can be used to remove larger-sized particles, but smaller, nanoparticles will most likely be collected by filtration (see next section, Air Filtration).

Air Filtration

Air filtration removes unwanted particulate from an air stream. Particulate air filters are classified as either mechanical or electrostatic filters. Although the two types of filters have important performance differences between them, both are fibrous media or membranes and are used extensively in HVAC and industrial applications. Efficiency is dependent on several factors including fiber diameters, packing density, and material used. A fibrous filter is an assembly of fibers that are randomly laid perpendicular to the airflow. The fibers may range in size from less than 1 μm to greater than 50 μm in diameter. Filter packing density ranges from 1%–30%. Fibers are made from cotton, fiberglass, polyester, polypropylene, or a number of other materials [Davies 1977]. Fibrous filters of different designs are used for various applications. Three types are used for capturing particulate:

- Flat-panel filters contain all the media in the same plane. This design keeps the filter face velocity and the filter media velocity roughly the same.

- Pleated filters have additional filter media added to reduce the air velocity through the filter. This allows for an increased collection efficiency for a given pressure drop. Alternatively, pleated filters can be used to reduce the pressure drop for a given airflow velocity because of the larger filter area.

- Pocket or baghouse filters allow the flow of exhaust air through small pockets or bags consisting of filter media. As with pleated filters, the increased surface area of the pocket filter reduces the velocity of the airflow through the filter media, allowing increased collection efficiency for small particles at a given pressure drop.

Figure 5 presents four different collection mechanisms that govern particulate air filter performance: f

- Diffusion is the result of the random (Brownian) motion of a particle. The particle may contact a fiber on its path through the filter. Interception occurs when the radius of a particle moving along an air streamline is greater than the distance from the streamline to the surface, thus causing the particle surface to contact the surface of the fiber. The particle adheres to the fiber due to intermolecular forces.

- Inertial impaction occurs when an air stream bends around a fiber, and a particle traveling in that air stream continues in a straight path due to particle inertia. The particle collides with the fiber and adheres to it due to intermolecular forces.

- Electrostatic attraction occurs when the particle and the fiber are oppositely charged. As the force of this attraction is governed by the charge-to-mass ratio of the particle, it becomes more effective as particle size decreases.

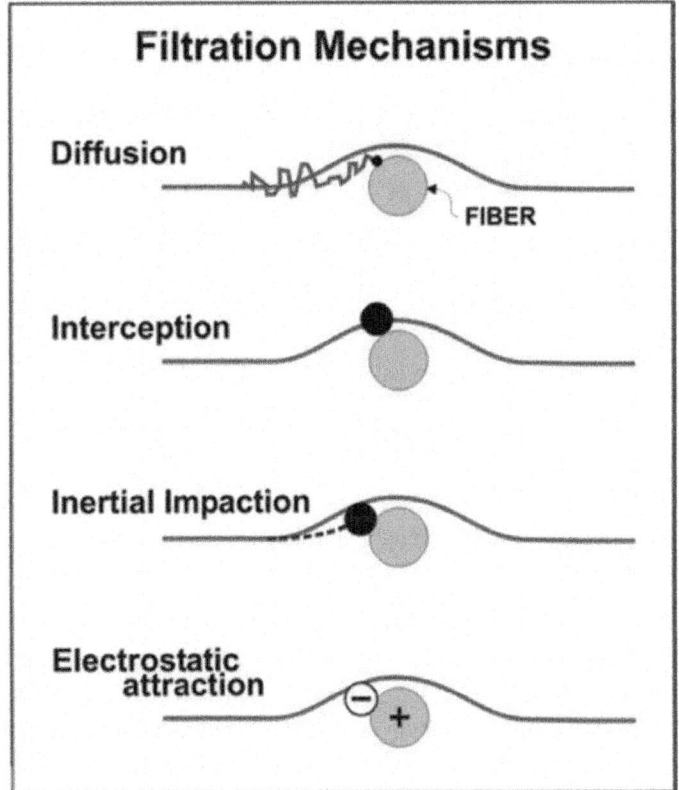

Figure 5: Four primary filter collection mechanisms

These mechanisms apply mainly to mechanical filters and are influenced by particle size. Impaction and interception are the dominant collection mechanisms for particles greater than 0.2 μm, and diffusion and electrostatic attraction are dominant for particles less than 0.2 μm, including nanomaterials. The combined effect of these collection mechanisms results in the classic collection efficiency curve, shown in Figure 6.

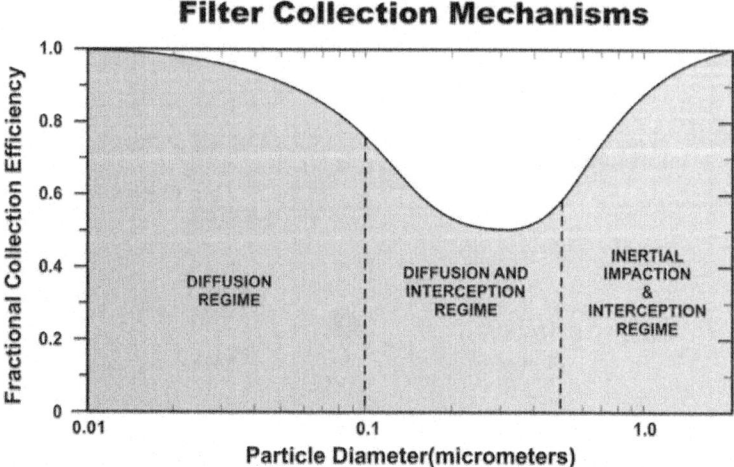

Figure 6: Collection efficiency curve, i.e., fractional collection efficiency versus particle diameter for a typical filter (Used with permission from Lee and Liu [1980].)

Research on common air filter materials has shown that fractional efficiency for collection of particles of different sizes is consistent with the single fiber theory [Heim et al. 2005; Kim et al. 2007; Shin et al. 2008]. Kim et al. [2006] found that humidity has little effect on particle collection efficiency. Huang et al. [2007b] determined that the use of electrostatic filters (commonly used for respirators) improves particle collection in the 0.1–1-μm particle size range. Testing of respirator filters showed that the most penetrating particle size (MPPS) shifted from 30–60 nm to 200–300 nm following treatment of respirators by liquid isopropanol, which removes electrostatic charges on the filter materials [Rengasamy et al. 2009]. This result suggests that capture by electrostatic forces is important for particles in the 250–300-μm range. Overall, filters appear to behave in a manner consistent with theoretical predictions that common filter materials allow for efficient collection through diffusion of nanoparticles less than about 10 nm [Heim et al. 2005; Huang et al. 2007b; Kim et al. 2007; Shin et al. 2008]. Some researchers have found evidence of thermal rebound, which increases particle penetration through filters for nanoparticles in the size range of 1–10 nm [Bałazy et al. 2004; Kim et al. 2006]; however, several other filter testing studies did not reveal this effect, even at higher temperatures [Heim et al. 2005; Huang et al. 2007b; Kim et al. 2007; Shin et al. 2008]. The thermal rebound effect is a result of the thermal velocity of the particle exceeding the critical sticking velocity for a particle on a filter, allowing the particle to move past the filter fiber and penetrate the filter. The critical sticking velocity of an incident particle is defined as the maximum impact speed at which the particle will stick to a surface; above this velocity, the particle will bounce and not stick to the filter. The primary adhesive forces for nanomatersized particles are the London-van der Waals forces. These forces are caused by random movement of electrons creating complementary dipoles between particle and filter material [Hinds 1999]. As the particle gets smaller, it is more difficult to remove the particle from surfaces. High efficiency particulate air (HEPA) filtration is commonly used for applications requiring reliably high filtration. By definition, HEPA filters are 99.97% efficient at the most penetrating particle size of 0.3 microns (Figure 6). These filters are disposable and are usually replaced when the pressure drop exceeds a predetermined number, typically 100 mm water gauge (wg). When properly sized and installed, HEPA filtration is appropriate for nanomaterial applications both for ventilation systems and respiratory protection.

Nonventilation Engineering Controls

Nonventilation engineering controls cover a range of control measures (e.g., guards and barricades, material treatment, or additives). Nonventilation controls can be used in conjunction with ventilation measures to provide an enhanced level of protection for nanomaterial workers. A variety of dust control methods have been used and evaluated in many industries and may be applicable to the processes used in the manufacturing and processing of nanomaterials [Smandych et al. 1998]. These methods include the enclosure of material-conveying equipment, such as belt and screw conveyers, as well as the use of pneumatic conveyance systems. Other work practices have been used to reduce the aerosolization of dust during bag filling, including minimizing leak paths by securing the bag to the outlet spout and wetting the outside of the bag to prevent surface dust from becoming airborne. Research over the years in a variety of industrial settings has shown that water spray application is effective in lowering respirable dust levels [Mukherjee et al. 1986]. The use of atomization nozzles was shown to be one of the most effective water-spray delivery systems in dust knockdown performance tests. Water sprays lower respirable dust concentrations by knocking down the dust, fibers, and particles, and they also can induce airflow to direct the remaining dust away from the workers. Other nonventilation engineering controls include many devices developed for the pharmaceutical industry, including isolation containment systems [Hirst et al. 2002]. One of the most common flexible isolation systems is glove box containment, which can be used as an enclosure around small-scale powder processes, such as mixing and drying. Rigid glove box isolation units also provide a method for isolating the worker from the process and are often used for medium-scale operations involving transfer of powders. Glove bags are similar to rigid glove boxes, but they are flexible and disposable. They are used for small operations for containment or protection from contamination. Another nonventilation control used in this industry is the continuous liner system, which allows the filling of product containers while enclosing the material in a polypropylene bag. This system is often used for off-loading materials when the powders are to be packed into drums.

Administrative Controls

Administrative controls and PPE are frequently used with existing processes where hazards are not well controlled. This could occur when engineering control measures are not feasible or do not reduce exposures to an acceptable level. Administrative controls (which include training, job rotation, work scheduling, and other strategies to reduce exposure) and PPE programs may be less expensive to establish but, over the long term, can be very costly to sustain. These methods for protecting workers have also proven to be less effective than other measures and often require significant effort by the affected workers [ACGIH 2013; DiNardi 2003]. A valuable application of administrative controls is as a redundancy to engineering controls. While the engineering controls provide the primary protection for the worker, the administrative controls serve as back-up should the engineering control fail. NIOSH recommends that facilities implement the following work practices as part of an overall strategy to control worker exposure to nanomaterials: (1) Educate workers on the safe handling of engineered nanomaterials to minimize the likelihood of inhalation exposure and skin contact. (2) Provide information on the hazardous properties of the materials being handled with instructions on how to prevent exposure. (3) Encourage workers to use handwashing facilities before eating, smoking, or leaving the worksite. (4) Provide additional control measures (e.g., use of a buffer area, decontamination facilities for workers if warranted by the hazard) to ensure that engineered nanomaterials are not transported outside of the work area. (5) Where there is the potential for area or personnel contamination, provide facilities for showering and changing clothes to prevent the inadvertent contamination of other areas (including take-home) caused by the transfer of nanomaterials on clothing and skin. (6) Avoid handling nanomaterials in the open air in a "free particle" state. (7) Store dispersible nanomaterials, whether suspended in liquids or in a dry particle form, in closed (tightly sealed) containers whenever possible. (8) Ensure work areas and designated equipment (e.g., balance) are cleaned at the end of each work shift, at a minimum, using either a HEPA-filtered vacuum cleaner or wet wiping methods (where the use of liquid does not create additional safety hazards). Dry sweeping (i.e., using a broom) or compressed air should not be used to clean work areas. Cleanup should be conducted in a manner that prevents worker contact with wastes. (9) Dispose of all waste material in compliance with all applicable federal, state, and local regulations. (10) Avoid storing and consuming food or beverages in workplaces where nanomaterials are handled [NIOSH 2009a].

Personal Protective Equipment (PPE)

PPE (e.g., respirators, gloves, protective clothing) is the least desired option for controlling worker exposures to hazardous substances. PPE is used when engineering and administrative controls are not feasible or effective in reducing exposures to acceptable levels or while controls are being implemented. It is the last line of defense after engineering controls, work practices, and administrative controls. A program that addresses the hazards present, employee training, and PPE selection, use, and maintenance should be in place when PPE is used.

Skin Protection

Nanomaterials have been shown to accumulate in hair follicles, and quantum dots have been shown to penetrate the skin into the dermis [Smijs and Bouwstra 2010]. Flexing the skin may enhance skin penetration [Smijs and Bouwstra 2010; Tinkle et al. 2003]. Woskie [2010] recommends wearing gloves, gauntlets, and laboratory clothing or coats when working with nanoparticles. Other studies of specifically engineered nanomaterials have resulted in the material not penetrating beyond the stratum corneum. Of importance is to establish a barrier between the potentially hazardous material and the skin. Airtight polyethylene was found to be more resistant to nanoparticle penetration by diffusion than cotton or polyester; gloves made of latex, neoprene, or nitrile resisted nanoparticle penetration "during exposure of a few minutes" [Woskie 2010]. Proper selection of gloves should take into account the resistance of the glove to the nanomaterial and any other chemicals or liquids with which the hands may come into contact. Gloves should be changed whenever they show visible signs of wear or contamination. Gao et al. [2011] studied nano- and submicron-size (30–500 nm) iron oxide particle penetration through some protective clothing materials. They found that particle penetration increased with increasing wind velocity and increasing particle size. Results from the study indicated that the MPPS for protective clothing materials tested was found to be about 300 to 500 nm, compared to an MPPS for N-95 respirators of 50 nm.

Respiratory Protection

Respiratory protection is used to reduce worker exposures to acceptable levels in the absence of effective engineering controls, during the installation or maintenance of engineering controls, for short-duration tasks that make engineering controls impractical, and during emergencies. The decision to use respiratory protection should be based upon professional judgment, hazard assessment, and risk management practices to keep worker inhalation exposures below an internal control or an exposure limit. Several types of NIOSH-certified respirators (e.g., disposable filtering facepiece, half-mask elastomeric, full facepiece, powered, airline, self-contained) can provide different levels of expected protection to airborne particulate when used in the context of a complete respirator program [60 Fed. Reg. 30336 (1995); NIOSH 2004]. In a survey designed to better understand health and safety practices in the carbonaceous nanomaterial industry, NIOSH found half-mask elastomeric particulate respirators fitted with HEPA filtration media to be the most commonly used respiratory protection, followed by disposable filtering facepiece respirators [Dahm et al. 2011]. The 2009 NIOSH Approaches to Safe Nanotechnology document as well as the Current Intelligence Bulletins on titanium dioxide and carbon nanotubes contain recommendations on respirator use and selection when working with nanoparticles. Recommendations from other organizations and a discussion of the scientific rationale for respirator selection have been reviewed [Shaffer and Rengasamy 2009]. Current respirator performance research suggests that NIOSH's traditional respirator selection tools apply to nanoparticles. NIOSH-certified respirators should provide the expected levels of protection, consistent with their assigned protection factor, and should be selected according to the NIOSH Respirator Selection Logic [NIOSH 2004] by the person who is in charge of the program and knowledgeable about the workplace and the limitations associated with each type of respirator. As part of the risk assessment process, respirators with 95-, 99-, or 100-class filters can be selected for workplaces with concentrations of nanoparticles near their MPPS (50 to 100 nm). Furthermore, NIOSH recommends that all elements of the OSHA Respiratory Protection Standard [29 CFR 1910.134] for both voluntary and required respirator use should be followed [63 Fed. Reg. 1152 (1998)]. Selection of respiratory protection for airborne particulate contaminants is typically done by dividing the measured or anticipated time-weighted average concentration of the airborne contaminant by the OEL and comparing that quotient to the respirator's assigned protection factor (APF). Alternatively, the respirator's APF can be multiplied by the OEL to find its maximum use concentration (MUC). The MUC is then compared to the TWA to select the appropriate respirator. In the absence of an

OEL for nanoparticles, Woskie [2010] recommends that a health and safety professional "familiar with the workplace" choose the appropriate respirator based on goals for nanoparticle control, sampling results, and the capabilities of each type of respirator. The NIOSH respirator selection logic recommends (and it is mandated by OSHA where the use of respirators is required) that respirators in the workplace be used as part of a comprehensive respiratory protection program. The program should include written standard operating procedures; workplace monitoring; hazard-based selection; fit-testing and training of the user; procedures for cleaning, disinfection, maintenance, and storage of reusable respirators; respirator inspection and program evaluation; medical qualification of the user; and the use of NIOSH-certified respirators [NIOSH 2004]. Several studies have been conducted of respirator media filtration performance against nanoparticles. Many employers provide filtering facepiece respirators (FFRs) due to their common availability and low cost. One study of N95 FFRs showed penetration levels by nanoparticles in the size range of ~30 to 70 nm, which exceeded the 5% level allowed by NIOSH [Balazy et al. 2006]. A later study used two test methods (challenges using a monodisperse aerosol and a polydisperse aerosol similar to the NIOSH certification test) and compared particle penetration of N-95 FFRs [Rengasamy et al. 2007]. Those authors found that a monodisperse aerosol challenge test using particles from 20 nm to 400 nm in diameter resulted in a MPPS near 40 nm. The monodisperse test found that two respirators exceeded the NIOSH 5% allowed penetration, but the exceedance was not statistically significant, while the polydisperse challenge produced penetration levels from 0.61% to 1.24%. The NIOSH-allowed penetration level of < 0.03% was not exceeded by P100 FFRs, but two nanoparticle test aerosols did exceed 1% penetration for two N99 FFRs [Eninger et al. 2008; Rengasamy et al. 2009]. Five models of N95 and two models of P100 FFRs challenged with 4–30-nm monodisperse aerosols provided approved levels of protection [Rengasamy et al. 2008]. Rengasamy et al. [2009] tested two models each of N95 and P100 respirators with monodisperse aerosols in the 4–30-nm range and the 20–400-nm range. The penetration levels were less than the NIOSH-allowed levels of < 5% and < 0.03% across all test methods used. The penetration was < 4.28% for the N95 respirators and < 0.009% for the P100 respirators at the MPPS range of 30–60 nm. NIOSH-certified FFRs have been shown to provide "expected levels of filtration efficiency against polydisperse and monodisperse aerosols > 20 nm in size" [Rengasamy and Eimer 2011]. A study showed that eight commercially purchased models of NIOSH-approved N95 and P100 and CE-marked FFR models "provided expected levels of laboratory performance against nanoparticles" [Rengasamy et al. 2009].

NANOTECHNOLOGY PROCESSES AND ENGINEERING CONTROLS

Primary Nanotechnology

Production and Downstream Processes Currently, nanomaterials are produced using a variety of methods that provide conditions for the formation of desired shapes, sizes, and chemical composition. These production processes can be separated into six categories [HSE 2004; NNI, no date]:

- Gas phase processes, including flame pyrolysis, high-temperature evaporation, and plasma synthesis. This process involves the growth of nanoparticles by homogenous nucleation of supersaturated vapor. Nanoparticles are formed in a reactor at high temperatures when source material in solid, liquid, or gaseous form is injected into the reactor. These precursors are supersaturated by expansion and cooled prior to the initiation of nucleated growth. The size and composition of the final materials depend on the materials used and process parameters.

- Chemical vapor deposition (CVD). This process has been used to deposit thin films of silicon on semiconductor wafers. The chemical vapor is formed in a reactor by pyrolysis, reduction, oxidation, and nitridation and deposited as a film with the nucleation of a few atoms that coalesce into a continuous film. This process has been used to produce many nanomaterials including TiO_2 , zinc oxide, silicon carbide, and, possibly most importantly, CNTs. The use of fluidized bed technology has been adopted as a way to prepare CNTs on a large scale at low cost [Wang et al. 2002]. This technology fluidizes CNT agglomerates and produces high yields necessary for larger-scale operations.

- Colloidal or liquid phase methods. Chemical reactions in solvents lead to the formation of colloids. Solutions of different ions are mixed to produce insoluble precipitates. This method is a fairly simple and inexpensive way to produce nanoparticles and is often used for the synthesis of metals (e.g., gold, silver). These nanomaterials may remain in liquid suspension or may be processed into dry powder materials often by spray drying and collection through filtration. ƒ Mechanical processes including grinding, milling, and alloying. These processes create nanomaterials by a "top-down" method that reduces the size of larger bulk materials through the application of energy to break materials into smaller and smaller particles. This technique has been referred to as nanosizing or ultrafine grinding.

- Atomic and molecular beam epitaxy. Atomic layer epitaxy is the process of depositing monolayers (i.e., layers one molecule thick) of alternating materials and is commonly used in semiconductor fabrication. Molecular beam epitaxy is another process for depositing highly controlled crystalline layers onto a substrate.

- Dip pen lithography. A "bottom-up" method is a production process that involves depositing a chemical on the surface of a substrate using the tip of an atomic force microscope (AFM). The AFM tips are coated with the chemical, which is directly deposited on a substrate in a specific pattern.

Downstream processes use engineered nanomaterials for product application and development. Examples of these tasks or operations include weighing, dispersion/sonication, mixing, compounding/extrusion, electro-spinning, packaging, and maintenance. These activities should be evaluated for potential sources of exposure.

Engineering Control Approaches to Reducing Exposures

Engineering controls are used to remove a hazard or place a barrier between the worker and the hazard, and though costs of engineering controls may be higher than that of administrative controls or PPE initially, over the long term, operating costs are often lower. A major advantage of engineering controls is that, when properly designed, they require little or no user effort or training to be effective. Many industries have implemented engineering controls to reduce exposure and risk of disease among their workers. The pharmaceutical industry uses hazardous (i.e., biologically active) liquids and powders that often do not have OELs. To address these hazards, the pharmaceutical industry has adopted a performancebased strategy using exposure control limits. This approach is based on establishing qualitative or semiquantitative criteria for assessing risk associated with the compounds and matching that information with known exposure-control approaches [Naumann et al. 1996]. Many of the processes used in pharmaceutical production are similar to those used in the nanoparticle industries discussed above and include blending, mixing, and handling of hazardous compounds in liquid and powder form. The general control concepts required for working with hazardous materials include specification of general ventilation, LEV, maintenance, cleaning and disposal, PPE, IH monitoring, and medical surveillance [Naumann et al. 1996]. Particular work practices, such as using HEPA-filtered vacuums instead of dry sweeping, are required. In addition, routine IH and medical monitoring ensure that work practices and engineering controls are effective. Source containment is considered the highest level in the containment hierarchy and is used by the pharmaceutical industry [Brock 2009]. This category contains many options including elimination, substitution, product modifications, process modifications, and equipment modifications. These steps could include reworking the process to reduce the number of times material is transferred or keeping the product in solution to minimize aerosolization potential. The next level of control for capturing process emissions is the use of engineering controls such as glove boxes, downflow booths, and local exhaust ventilation. Genaidy et al. [2009] conducted a detailed hazard analysis of a CNF manufacturing process and suggested the following potential sources of workplace exposure to nanomaterials:

- Leakage and spillage from reactors and powder processing equipment
- Manually harvesting product from reactors

    Discharging product into containers

    Transporting containers of intermediate products to the next process

    Charging the powders into processing equipment

    Weighing out powder for shipment f Packaging material for shipment

    Storing material between operations

    Cleaning equipment to remove debris stuck to side walls

    Changing filters on dust collection systems and vacuum cleaners f Further processing of products containing nanomaterials (e.g., cutting, grinding, drilling)

This detailed analysis, along with the review of exposure assessment studies in nanomaterial production and downstream user facilities described below, identify common processes that may lead to worker exposure to nanomaterials. This section provides some information on engineering control approaches that may be applicable for these common processes/tasks. Table 1 shows a generic process list along with applicable engineering controls and references. The engineering control column provides a framework for identifying exposure controls for particular processes. The third column shows the industry in which these control approaches have been tested. References are listed in the fourth column for studies that apply to each of these processes and controls.

Table 1: Engineering controls and associated tasks for various industries

Process/task	Engineering control	Industry	Reference
Reactor fugitive emissions	Enclosure	Nanotechnology	Tsai et al. 2009b Lee et al. 2011
Product harvesting	Glovebox	Nanotechnology	Yeganeh et al. 2008
Reactor cleaning	Spot LEV system/fume extractor	Nanotechnology	Methner 2008
Small-scale weighing	Chemical fume hood	Nanotechnology	Tsai et al. 2009a Ahn et al. 2008 Tsai et al. 2010
	Biological safety cabinet	Nanotechnology and laboratory	Cena and Peters 2011 Macher and First 1984
	Glovebox isolator	Pharmaceutical	Walker 2002 Hirst et al. 2002
	Nano fume hood	Pharmaceutical	
	Air curtain isolation hood	Nanotechnology/research	Tsai et al. 2010
Product discharge/ bag filling	Discharge/collar hood	Silica and pharmaceutical	ACGIH 2013 HSE 2003e Hirst et al. 2002
	Continuous liner	Pharmaceutical	Hirst et al. 2002
	Inflatable seal	Pharmaceutical	Hirst et al. 2002
Bag/container emptying	Bag dump station	Silica	HSE 2003d Heitbrink and McKinnery 1986 Cecala et al. 1988
Large-scale weighing/ handling	Ventilated booth	Pharmaceutical	Hirst et al. 2002 Floura and Kremer 2008 HSE 2003b
Nanocomposite machining	High velocity-low volume	Woodworking	
	Wet suppression	Nanotechnology	Bello et al. 2009
Air filter change-out	Bag in-bag out	Pharmaceutical	

Ventilation and GeneralConsiderations

It is important to confirm that the LEV system is operating as designed by regularly measuring exhaust airflows. A standard measurement - hood static pressure - provides important information on the hood performance, because any change in airflow results in a change in hood static pressure. For hoods designed to prevent exposures to hazardous airborne contaminants, the ACGIH Industrial Ventilation: A Manual of Recommended Practice for Operation and Maintenance recommends the installation of a fixed hood static pressure gauge [ACGIH 2010].

In addition to routinely monitoring the hood static pressure, additional system checks should be completed periodically to ensure adequate system performance, including smoke tube testing, hood slot/face velocity measurements, and duct velocity measurements using an anemometer. A dry ice test is another method of evaluation designed to qualitatively determine the containment performance of fume hoods. These system evaluation tasks should become part of a routine preventative maintenance schedule to check system performance. It is important to note that the collection and release of air contaminants may be regulated; companies should contact agencies responsible for local air pollution control to ensure compliance with emissions requirements when implementing new or revised engineering controls. To reduce the risk of exposure to nanomaterials, a few standard precautions should be followed in areas where exposures may occur:

- Isolate rooms where nanomaterials are handled from the rest of the plant with walls, doors, or other barriers.

- Maintain production areas where nanomaterials are being produced or handled under negative air pressure relative to the rest of the plant.

- Install hood static pressure gauges (manometers) near hoods to provide a way to verify proper hood performance.

- When possible, place hoods away from doors, windows, air supply registers, and aisles to reduce the impact of cross drafts.

- Provide supply air to production rooms to replace most of the exhausted air.

- Direct exhaust air discharge stacks away from air intakes, doors, and windows. Consider environmental conditions, especially prevailing winds.

Exposure Control

Technologies for Common Processes In a review of exposure assessments conducted at nanotechnology plants and laboratories, Brouwer [2010] determined that activities that resulted in exposures included harvesting (e.g., scraping materials out of reactors), bagging, packaging, and reactor cleaning. Downstream activities that may release nanomaterials include bag dumping, manual transfer between processes, mixing or compounding, powder sifting, and machining of parts that contain nanomaterials. Particle concentrations during production activities ranged from about 103–105 particles/cm^3 . Most studies showed bimodal particle distributions with modes of about 200–400 nm and 1,000–20,000 nm, indicating that the emissions are dominated by aggregates and agglomerates. With the exception of leakage from reactors when primary manufactured nanoparticles may be released, workers are believed to be primarily exposed to agglomerates and aggregates. Methner et al. [2010] summarized the findings of exposure assessments conducted in 12 facilities with a variety of operations: seven were R&D labs, one produced CNTs, one produced nanoscale TiO_2 , one produced nanoscale metals and metal oxides, one produced silica-iron nanomaterials, and one manufactured nylon nanofibers. The most common processes observed at these facilities were weighing, mixing, collecting product, manua

Engineering controls used included portable vacuums with filters, laboratory fume hoods, portable LEV systems, ventilated walk-in enclosures, negative pressure rooms, and glove boxes. Tasks such as weighing, sonicating, and cleaning reactors showed evidence of nanomaterial emissions. The highest nanoparticle exposures measured occurred inside spray booth-type enclosures and during a spray dryer collection drum change-out. Other activities that resulted in higher exposures include reactor cleanout tasks (e.g., brushing and scraping slag material). Incidental (nonprocess) ultrafines were measured from a variety of sources, including electric arc welding, operating a propane-powered forklift, and the exhaust of a portable vacuum outfitted with filters. From a review of published studies, some common sources of nanoparticles and fine particles can be identified. As expected, those processes that require material handling resulted in worker exposure to nanomaterials. Other activities that require operator interface with the reactor can result in nanoparticle exposure, and background concentrations may increase as a result of leakage from reactors under positive pressure. In addition, several studies found that evaluation of process emissions and exposure should take into account major sources of incidental nanoparticles that may be present in the workplace and also sources of natural nanoparticles, e.g., tree pollen brought into the work area through the facility HVAC system. Common incidental sources include diesel exhausts in outdoor air, welding fumes, forklifts, and gasfired heaters. Several studies showed that the use of engineering controls can reduce operator exposure, while one study showed that a poorly designed enclosure actually increased exposure [Cena and Peters 2011; Methner et al. 2007; Tsai et al. 2009a, 2010; Yeganeh et al. 2008]. The following sections describe applicable engineering controls for common processes used by nanotechnology companies described in the literature. For each control, a background is given along with a summary of relevant research conducted on their performance. Many of the control concepts discussed in this section come from the HSE Control Guidance Sheets in COSHH Essentials: Easy Steps to Control Chemicals [HSE 2003a,b,c,d] and the ACGIH Industrial Ventilation Manual [ACGIH 2013]. Table 2 lists common processes and tasks, along with potential emission points and the section or figure(s) that address those processes.

Table 2: Process/tasks and emission

Process/task	Potential emission/ exposure points	See section	See figures
Production of bulk nanomaterials	Reactor fugitive emissions	3.4.1	7, 8
	Product harvesting	3.4.1	12
	Reactor cleaning	3.4.1	
Downstream processing	Product discharge/bag filling	3.4.3.1	14, 15, 16
	Bag/container emptying	3.4.3.2	17
	Small-scale weighing	3.4.2	10, 11, 12, 13
	Machining of products	3.4.3.4	
Product packaging	Small-scale weighing/handling	3.4.2	10, 11, 12, 13
	Large-scale weighing/handling	3.4.3.3	18
	Product packaging	3.4.3	14, 15, 16, 18
Maintenance	Facility equipment cleaning	3.4.4	
	Air filter change-out	3.4.4.1	19
	Spill clean-up	3.4.4.2	

Reactor Operation and Cleanout Processes

Harvesting material from reactors has been identified as a potentially high exposure activity in several manufacturing plants [Demou et al. 2008; Lee et al. 2010, 2011; Methner 2008; Yeganeh et al. 2008]. In addition, cleanout of reactors has contributed to increasing facility concentrations and exposures to operation and maintenance workers. Leakage from pressurized reactors can also contribute to background concentrations and result in exposure to employees throughout the facility. When the reactors are small, some facilities have placed them inside fume hoods to help control fugitive emissions. Two studies have shown that when the reactor is housed in a well-designed and operated fume hood, particle loss to the work environment is low [Tsai et al. 2009b; Yeganeh et al. 2008]. When the reactors are larger, enclosures can be built that isolate the reactor from the environment and seek to reduce fugitive emissions (Figure 7). Methner et al. [2010] summarized airborne measurements in 12 facilities that processed nanomaterials, including manufacturers and research and development labs. The authors found that some of the highest measured exposures occurred during reactor cleanout tasks, which included brushing and scraping slag material from the reactor walls and during torch cleaning. Demou et al. [2008] evaluated exposure to nanoparticles at a pilot-scale nanomaterial production facility. The major emission source was determined to be the production unit as the airborne particle concentrations rose when the unit was started and fell when production rate was decreased. The other task that resulted in substantial particle release was cleaning of the reactor using a vacuum cleaner not fitted with a HEPA filter. Evans et al. [2010] studied nanoparticle concentrations in a facility that manufactured and processed carbon nanofibers (CNFs). During the thermal treatment of the CNFs in a reactor under positive pressure, elevated concentrations of non-CNF ultrafines were released.

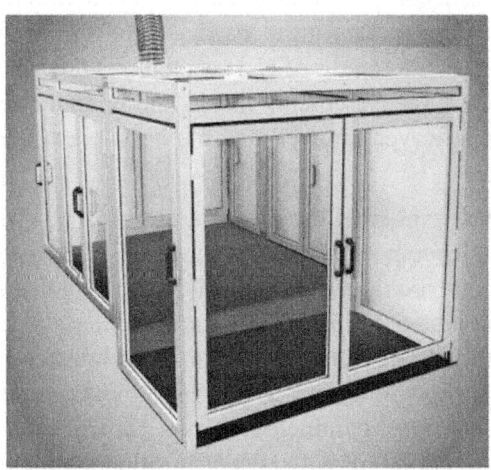

Figure 7: A large-scale ventilated reactor enclosure used to contain production furnaces to mitigate particle emissions in the workplace (Used with permission from Flow Sciences, Inc.)

Lee et al. [2010] conducted personal, area, and real-time sampling in seven CNT plants. Results showed that nanoparticles and fine particles were most frequently released upon opening the chemical vapor deposition (CVD) furnace. Catalyst preparation and the opening of the CVD furnace resulted in the release of nanoparticles in the range of 20–50 nm. Lee et al. [2011] also evaluated workplace exposures to nanoscale TiO2 at manufacturing plants. In one TiO_2 plant, the reactor was small and was placed in a fume hood; the entire process was conducted in that hood. Even though the reactor was located in the hood, high concentrations of nanoparticles were measured outside the hood. Worker exposure increased during product harvesting because the worker put his head into the hood to brush out the product powder. A second TiO_2 plant isolated the large-scale reactor with a vinyl curtain and used a glove box for the harvesting of product from the reactor. Overall, airborne particle concentrations were fairly stable during production although increases occurred during both the operation of a process vacuum pump and welding conducted in the facility. Yeganeh et al. [2008] evaluated a small facility producing carbonaceous nanomaterials including fullerenes. The process involved the production of materials in an arc furnace that was enclosed in a ventilated fume hood. This hood had a plastic front face shield and ports that allowed worker access during the process. The process involved placing graphite rods into the furnace, volatilizing the graphite in the furnace, producing raw soot, and using a scoop and brush to remove raw soot into a jar. At the beginning or end of each day, the reactors were completely cleaned by manual sweeping and vacuuming to remove residual soot. Real-time particle analyses showed that physical handling of material (sweeping of the reactor) resulted in the aerosolization of ultrafine particles. Measurements inside and outside the reactor enclosure (i.e., fume hood), however, showed that the hood was effective at containing particulates.

Methner [2008] evaluated the use of a portable LEV unit for controlling exposure during cleanout of a vapor deposition reactor used for producing nanoscale metal catalytic materials comprised of manganese, cobalt, or nickel. Following the automated collection of product materials, an operator cleaned out slag and waste product from the reactor using brushes and scrapers. Initial measurements had shown this task to be a high-exposure task for the operator. A follow-up survey was conducted at the facility using a commercially available fume extraction unit with HEPA filtration to pull airborne dusts away from the operator during cleanout. Analysis of real-time instrumentation and filter samples analyzed for metals showed an average reduction in airborne concentrations of 88%–96% during three cleanout procedures. Emission sources related to reactor operations, harvesting, and maintenance can be categorized as fugitive or task–based. The approaches that have been used for controlling fugitive emissions from the reactor have primarily been ventilated enclosures. Laboratory fume hoods and glove boxes can be used when the reactor is small, typical of R&D or pilot operations. Where the production reactors are larger, custom-fabricated enclosures often constructed from a polycarbonate, transparent thermoplastic material, or vinyl curtains have been used to reduce emissions (Figure 7). When designing these types of enclosures, it is necessary to consider reactor access needs, determination of exhaust airflows capable of maintaining a negative pressure (even during the opening of access doors), and accommodation of heat loads generated by the process. Failure of containment can result from not carefully addressing these key design needs. When looking at pressure differentials, it is important to study the airflow to minimize turbulent situations that can actually increase particle release rather than containing the particles.

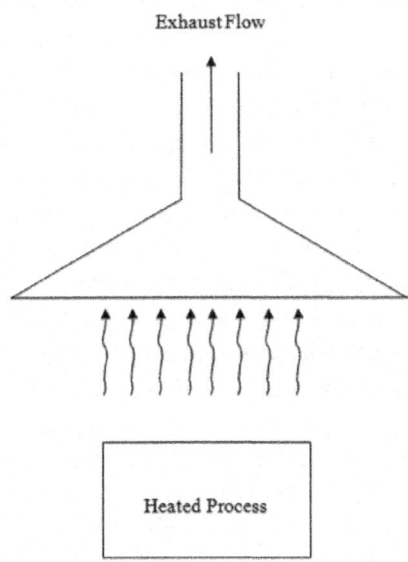

Figure 8: A canopy hood used to control emissions from hot processes

When a process is heated, the use of canopy hoods (Figure 8) may be another reasonable alternative as long as the design meets the operational and facility exposure control requirements [ACGIH 2013; McKernan and Ellenbecker 2007]. Even if the process does not involve heat, contaminant capture velocities suitable for gas/vapor contaminants (rather than coarse particulates) may be sufficient, as ultrafine and nanoparticles possess negligible inertia and follow the flow stream well. When controlling exposures during operations such as product harvesting and reactor cleanout, solutions such as spot LEV systems (e.g., a fume extractor) or containment may be acceptable alternatives. Manual harvesting of product materials may be better suited for higher-level enclosure controls such as a glove box or a specially designed enclosure to provide good capture while minimizing loss of product materials. The use of a commercially available fume extractor has been shown to be effective in reactor cleanout and provides a flexible solution that may meet facility needs across a range of operations [Methner 2008]. Selection of any control should be evaluated to ensure worker acceptance and use as well as verifying that it meets the exposure control objectives.

Small-scale Weighing and Handling of Nanopowders

Small-scale weighing and handling of nanopowders are common tasks; examples include working with a quality assurance/control sample and processing small quantities in downstream industries. During these operations, workers may weigh out a specific amount of nanomaterials to be added to a process such as mixing or compounding. The tasks of weighing out nanomaterials can lead to worker exposure primarily through the scooping, pouring, and dumping of these materials. Many different types of commercially available laboratory fume hoods can be employed to reduce exposure during the handling of nanopowders. Other controls have also been used in the pharmaceutical and nanotechnology industries for containment of powders during small-quantity handling and manipulation. They include glove boxes, glove bags, biological safety cabinets or cytotoxic safety cabinets, and homemade ventilated enclosures. Methner et al. [2007] evaluated a university-based research lab that used CNFs to produce high-performance polymer materials. Several processes were evaluated during the survey: chopping extruded materials containing CNFs, transferring and mixing CNFs with acetone, cutting composite materials, and manually sifting oven-dried CNFs on an open benchtop. Real-time monitoring did not identify any process as a substantial source of airborne CNF emissions; however, weighing/mixing of CNFs in an unventilated area resulted in elevated particle concentrations compared to background. Other studies have shown that benchtop activities such as probe sonication of nanomaterials in solution can also result in emission of airborne particles [Johnson et al. 2010; Lee et al. 2010]. Producing dispersions by sonication is a primary operational step, and the industrial hygiene assessment should address the sound level exposure as well as the potential exposure to aerosols of nanomaterials from the sonication. Maintaining the sonicator/dispersion process within an enclosure such as a hood can be an effective means for mitigating the noise and aerosol exposure.

Fume Hood Enclosures

In 2006, a survey was conducted of international nanotechnology firms and research laboratories that reported manufacturing, handling, researching, or using nanomaterials [Conti et al. 2008]. All organizations participating in the survey reported using some type of engineering control. The most common exposure control used was the traditional laboratory fume hood with two-thirds of firms reporting the use of a fume hood to reduce exposure to workers. These devices have been used for many years in research laboratories to protect workers from chemical and biological hazards. The design and operation of the fume hood is an important factor when considering good exposure control. Traditional designs for laboratory fume hoods create airflow patterns that form recirculation regions inside the hood. In addition, airflow around the worker, as shown in Figure 9, creates a negative pressure region downstream of the worker, which may provide a mechanism for the transport of materials out of the hood as well as into the breathing zone of the worker. Recent research has shown that the laboratory fume hood may allow the release of nanomaterials during their handling and manipulation [Tsai et al. 2009a]. This research evaluated exposures related to the handling (i.e., scooping and pouring) of powder nanoalumina and nanosilver in a constant air volume (CAV) hood, a bypass hood, and a variable air volume (VAV) hood. This study showed that the CAV fume hood, in which face velocity varies inversely with sash height, allowed the release of significant amounts of nanoparticles during pouring and transferring activities involving nanoalumina.

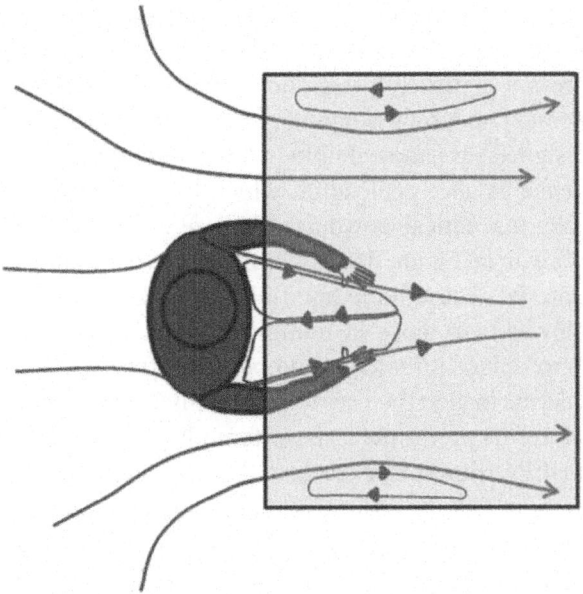

Figure 9: Schematic illustration of how wakes caused by the human body can cause transport of air contaminants into the worker's breathing zone

The particles that escaped the fume hood were circulated to the general room air and were not cleared by the general ventilation system for 1/2–2 hours. Sash heights both above and below the recommended height (corresponding to a face velocity of 80–120 ft/min) may lead to increased potential exposure for the user. In contrast, more modern hoods such as the VAV hood, which is designed to maintain the hood face velocity in a desired range regardless of sash height, yielded better containment of nanoparticles than the other hoods tested.

A meta-analysis of fume hood containment studies was conducted to identify the important factors that affect the performance of a laboratory fume hood [Ahn et al. 2008]. An analysis of factors affecting the containment performance of the hoods showed that worker exposures to air contaminants can be greatly impacted by a variety of operational issues. Increasing the distance between the contaminant source and the breathing zone leads to reduced exposure. Exposures can also be reduced by limiting the height/area of the sash opening; increasing the height of the sash opening increased the risk of hood containment failure. The presence of a manikin/human subject in front of the hood caused the greatest risk of hood failure among factors studied. This indicates that containment testing should include an operator or manikin to adequately assess hood performance. Face velocity did not make a significant difference in hood performance unless it was extremely high or low (> 150 ft/min or < 60 ft/min). Several hood operating factors showed an effect but were not statistically significant, including sash movement, hand and arm movement, pouring/weighing, and thermal load. New fume hoods specifically designed for nanotechnology are being developed primarily based on low-turbulence balance enclosures, which were initially developed for the weighing of pharmaceutical powders. The use of bench-mounted weighing enclosures, as seen in Figure 10, is common for the manipulation of small amounts of material. These fume hood-like LEV devices typically operate at airflow rates lower than those in traditional fume hoods and use airfoils at enclosure sills to reduce turbulence and potential for leakage. They also have face velocity alarms to alert the user to potentially unsafe operating conditions. Based on the hazards assessment, these fume hood-like LEV devices can be outfitted with HEPA filtration or connected to the ventilation exhaust system.

Biological Safety Cabinets The Centers for Disease Control and Prevention (CDC) divides biological safety cabinets (BSCs) into three classes: Class I, Class II, and Class III. The Class II BSCs are further divided into four subcategories (A1, A2, B1, B2) [DHHS 2009]. These hoods are used for processes that require operator and product protection. The BSC pulls air into the hood to protect the operator while providing a downward flow of HEPA-filtered air inside the cabinet to minimize cross-contamination along the work surface (see Figure 11). The most common BSC (Type II/A2) uses a fan to provide a curtain of HEPA-filtered air over the work surface. The downward moving air curtain splits as it approaches the work surface; some of the air is drawn to the front exhaust grille and the remainder to the rear grille. The air is then drawn back up to the top of the cabinet where it is recirculated or exhausted from the cabinet. In general, 70% of the air is HEPA-filtered and recirculated while 30% is filtered and then exhausted from the cabinet. The make-up air is drawn through the front of the cabinet. The air being drawn in acts as a barrier to protect the workers from contaminated air leaking out of the hood.

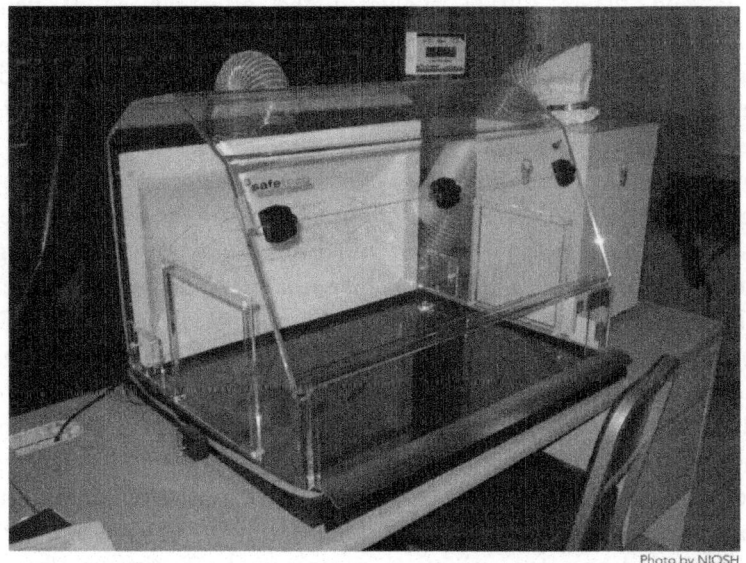

Figure 10: Nano containment hood adapted from a pharmaceutical balance enclosure

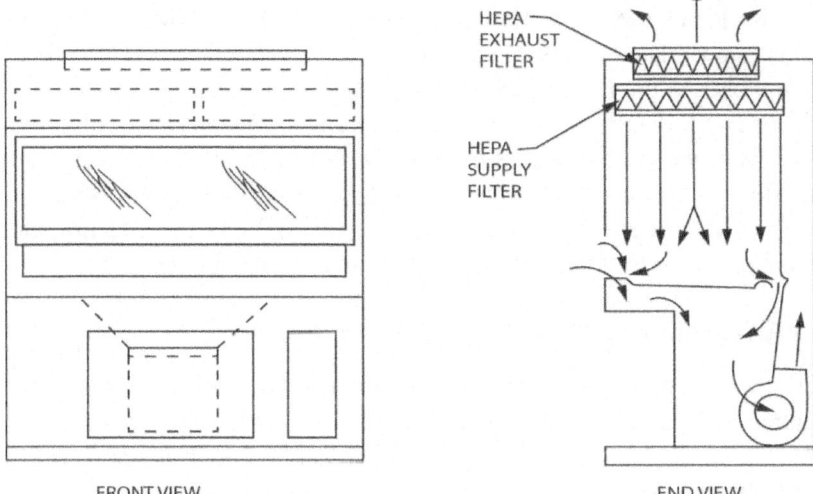

FRONT VIEW   END VIEW

Class II type

Figure 11: A tabletop model of a Class II, Type A2 biological safety cabinet (BSC) (Used with permission from ASHRAE [2011].)

Cena and Peters [2011] evaluated the effectiveness of ventilated enclosures including a Class II, Type A2 BSC and a custom fume hood during the manual sanding of epoxy test samples reinforced with CNTs. Sanding of CNT-epoxy materials released respirable-sized (micronsized) particles but generally no nano-sized particles. The respirable mass concentration in the operator's breathing zone while using the BSC was approximately two orders of magnitude lower than the concentration when using the custom fume hood. The use of the custom fume hood resulted in an increase of breathing zone concentrations of about one order of magnitude compared to the use of no controls. The custom fume hood had a low average face velocity of about 45 ft/min with high variability across the hood face. The authors suggested that the poor performance of the custom fume hood may have been due to its rudimentary design, which did not include a front sash or rear baffles. The lack of these common fume hood features along with the low average face velocity may have resulted in poor airflow distribution across the face and increased leakage. Macher and First [1984] evaluated the effect of airflow rates and operator activity on containment effectiveness for a Class II, Type B1 biological safety cabinet using bacterial spores released by two 6-jet collison nebulizers. The hood sash height correlated negatively with the containment effectiveness; that is, the higher of two sash heights provided better containment of the aerosol. In addition, working in the front half of the cabinet provided better protection than working in the rear half of the cabinet. The authors postulated that working in the rear of the hood caused the operator to move closer to the hood opening, blocking the opening and causing more turbulence and leakage around the sides of the hood. The operator withdrawing his arms from the hood caused significantly more leakage than moving arms side to side within the hood. The authors concluded that testing BSCs with persons working at them provides more information than static testing alone and that even well-designed cabinets lose a small fraction of aerosols.

Glove Box Isolators

A glove box isolator fully isolates (contains) a small-scale process and is sometimes referred to as a primary protection device (Figure 12) [HSE 2003a]. The design can be either the more typical hard unit or a soft, flexible containment unit (often referred to as a glove bag). Glove boxes provide a high degree of operator protection but at a cost of limited mobility and size of operation. In addition, cleaning the glove box may be difficult, and, to prevent exposures, operators should use caution when transferring materials and equipment into and out of the glove box. In general, glove boxes include a pass-through port, which allows the user to move equipment or supplies into and out of the enclosure.

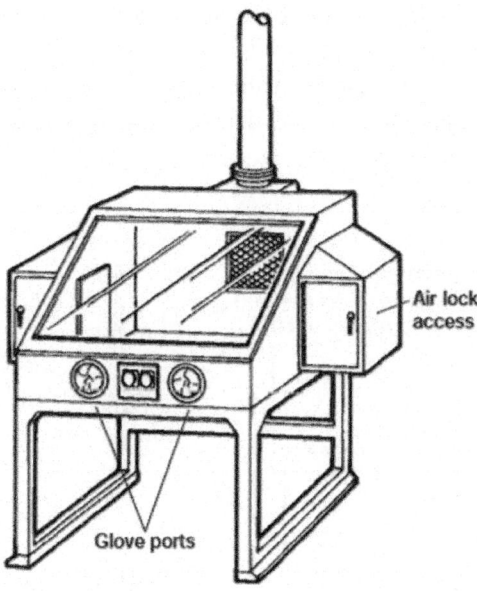

Figure 12: A glove box isolator for handling substances that require a high level of containment (Contains public sector information published by the Health and Safety Executive and licensed under the Open Government License v1.0.)

The performance of a glove box containment system was evaluated during weighing activities of fine lactose powder (a common pharmaceutical surrogate test material). Air samples were collected at four locations: inside the glove box, in the pass-through, in front of the glove box, and at the exit of the recirculating HEPA filter [Walker 2002]. The results of sampling a 10-minute task showed the average concentration measured inside the glove box was 298 µg/ m^3 , the average concentration in the integral pass-through was 35 µg/m^3 , and concentrations measured in the room, including downstream of the glove box exhaust, were below the analytical limit of detection of 1 µg/m^3 . Sample swabs of interior surfaces showed dust contamination within both the main glove box and pass-through. These results indicated that, although internal surfaces were contaminated with the materials, no leakage from the glove box was detected.

Air Curtain Fume Hood

A recent fume hood design addresses the known issues surrounding the recirculating flow patterns both inside the fume hood and around the operator (Figure 9). The air curtainisolated fume hood, as shown in Figure 13, uses a push-pull ventilation configuration created by a narrow planar jet from the sash to an exhaust slot along the base of the hood opening. Tsai et al. [2010] evaluated the performance of this hood during handling and manipulation of nanoparticles. In this test, measurements in the worker's breathing zone were taken while nanoalumina powders were manually transferred or poured between several 400-ml beakers.

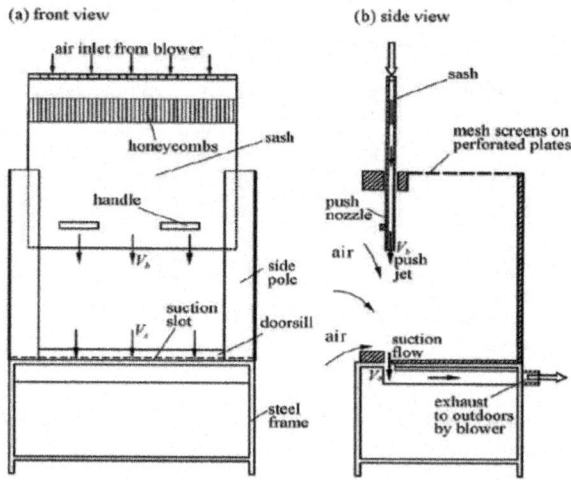

Figure 13: Air curtain safety cabinet hood that uses push-pull ventilation (Used with permission from Huang et al. [2007a].)

The air-curtain hood had very low particle release during all tested conditions (i.e., varying sash heights) with low but measurable release occurring at the lowest sash position. This same study showed that the particle leakage from two traditional fume hoods (both a CAV and VAV hood) exhibited substantial particle release during similar nanomaterial handling operations. This study suggested that the air curtain isolated hood may provide better containment performance during typical handling procedures.

Intermediate and Finishing

Processes Exposures resulting from the manual handling of powdered materials are common in industry. Reduction in worker exposure through implementation of careful work practices and appropriate engineering controls would benefit these operations. Dumping bags of powdered materials has been commonly reported in the literature for production and processing. Typically, a worker dumps the ingredients for one process into a hopper and then compacts or disposes of the empty bags. Ventilated bag-dumping stations have been used successfully in a variety of industries and applications.

The transfer of large quantities of nanomaterials requires different solutions adaptcd to the particular process. However, a few controls that are applicable to these common processes are available and have been evaluated for similar industrial operations. After the completion of production, many nanomaterials are sent for further processing. The powder product may be refined through a common process such as spray drying [Lindeløv and Wahlberg 2009]. Other studies have documented collection of fugitive emissions of nanomaterials from the process reactors using devices such as baghouse air filters [Evans et al. 2010]. In both of these operations, the nanomaterials are collected in a barrel or drum following the completion of these production steps. Several examples of engineered drum or bag filling solutions have been described elsewhere and could be implemented to reduce such releases [ACGIH 2013; Hirst et al. 2002]. These engineering controls consist of enclosing the product off-loading process by temporarily sealing the drum/bag to the filling vessel above and/or overbagging through a continuous liner type bagging system. The addition of a local exhaust ventilation hood near the drum/bag opening could also be used to capture airborne nanomaterials. Evans et al. [2010] studied nanoparticle concentrations in a facility that manufactured and processed carbon nanofibers (CNFs). The authors discussed four discrete events that resulted in elevations in airborne particle concentrations. The largest increases in particle concentration measured within the plant were related to manual handling processes, such as dumping product into lined drums and manual change-out and closing bags of final treated CNF product. Increases in particle concentrations were the result of the change-out and closing of the collection bag containing approximately 15 lbs of CNFs. Emissions from this event were almost entirely due to aerosolized CNFs. Tamping of the bag to settle contents (so that it could be adequately closed) and subsequent closing appeared to efficiently aerosolize CNF material through the bag opening into the workplace environment. This resulted in an increase in respirable mass concentration and a dark visible airborne CNF plume. A few studies have been conducted to look at the emission of nanoparticles from downstream products during machining of nanocomposites. Methner et al. [2007] reported increases in total carbon (a marker for nanoparticles), particle number, and mass concentration during the wet sawing of a CNF-impregnated composite. However, the increase in particle concentration was primarily of particles greater than 400 nm in diameter. Vorbau et. al. [2009] evaluated nanoparticle release from oak and steel panels coated with polyurethane mixed with zinc oxide nanoparticles. A standard abrasive test rig was used to provide uniform conditions for testing the release of particles from the surface of the panels. During the abrasion tests, no significant release of particles below 100 nm was observed. However, the nanoscale zinc oxide particles were embedded in the aerosols with larger surface area. Bello et al. [2009] evaluated the release of nanoscale particles during dry and wet cutting of nanocomposite materials. Two composites were used for evaluation: a CNT-enhanced graphite prepreg laminate sheet and a woven alumina fiber cloth with CNTs grown on the surface of the fibers. Significant exposures to nanoscale particles were generated during dry cutting of all composites with emission levels being related to composite material and thickness; wet cutting reduced exposures to background levels. For all processes/tasks discussed, engineering controls should be adapted for the specific process. Acceptable exhaust volumes and capture velocities may differ from currently available guidance due to differences in materials being handled. Pilot testing of any controls should be conducted to evaluate proper control operation and verify that exposures are controlled to desired levels.

Product Discharge/Bag Filling

The process of filling bags with nanomaterials is commonly done following large-scale production or refining processes. The off-loading of product after spray drying, for example, may be a significant source of exposure when post-processing nanomaterials. In the spray-drying process, a mixture of liquid and powder ingredients (slurry) is sprayed within a large sealed tank. Heat within the tank dries the slurry droplets, leaving a powder as the finished product. When the process is completed, the powder product is commonly discharged into a bulk tote or drum before packaging. Methner et al. [2010] reported exposure measurements at 12 facilities and noted that the highest backgroundadjusted concentration was observed during spray dryer drum changeout. Evans et al. [2010] reported exposures related to changing out a drum that collected fugitive CNF materials from a process reactor using a baghouse filtration system. Even though the processes differ, the tasks for each of these steps are similar and include the removal of the drum from the process outlet. These drums are often sealed to the process outlet minimizing exposure during production but potentially expose workers when removing the drums or liners.

A ventilated, collar-type hood around the discharge point can help minimize worker exposure to dust. Figure 14 presents a control approach for filling bags with solid powder materials [HSE 2003c]. The control includes the specification of a ventilated enclosure around the powder discharge outlet and applies to filling smaller product bags as well as intermediate bulk containers. This design guidance recommends an inward air velocity of 200 ft/min (1.0 m/s) into the enclosure.

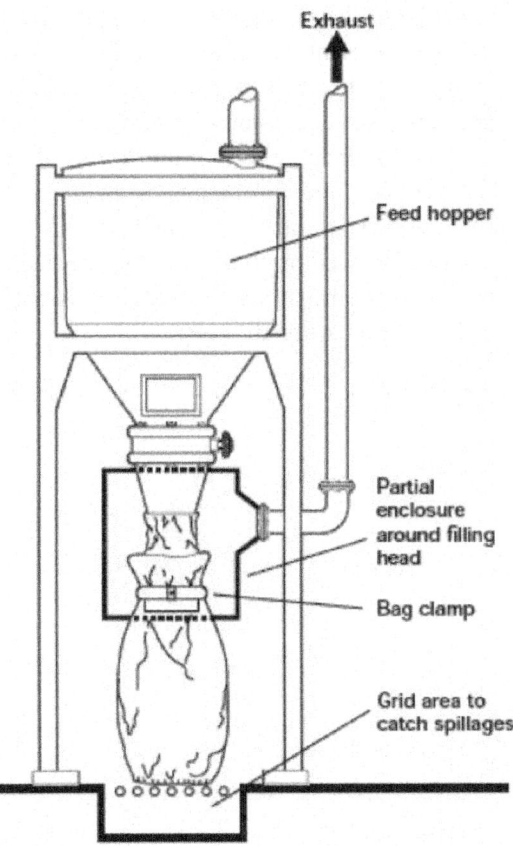

Figure 14. Ventilated collar-type exhaust hoods for containing dust during product discharge or manual bag filling (Contains public sector information published by the Health and Safety Executive and licensed under the Open Government License v1.0.)

The ACGIH Industrial Ventilation Manual [ACGIH 2013], Design plate VS-15-02, Bag Filling, is similar in design to the HSE exhaust hood (Figure 14) but specifies an overall hood flow rate of 400–500 ft^3 /min for nontoxic dust or 1,000–1,500 ft3 / min for toxic dust with a maximum inward air velocity of 500 ft/min. These flow rates have been specified for common industrial powders and may need to be adjusted based on the process and properties of the nanoscale materials being addressed to prevent excessive loss of product.

In addition to ventilation solutions, other dust control approaches have been used in a variety of industries and should be applicable for nanomaterial production. For example, an inflatable seal can be used to create a dust-tight seal on the discharge outlet of a spray dryer during the product discharge/bag filling process (Figure 15). The seal inflates during the product transfer from the process to the packaging bag (providing the seal) and deflates once the transfer is completed to allow removal of the bags. These systems are available on many commercially available bulk bag filling systems [Hirst et al. 2002]. Another system that can be used to contain powders during process off-loading/ emptying is the continuous liner system (Figure 16). Polypropylene liners are often used when products are discharged from the industrial processes into the intermediate or final product containers. In this operation, a sleeve of polypropylene liners is stowed around the circumference of the discharge outlet. The first liner, the bottom having been sealed, is pulled down into the overpack (usually a drum or a cardboard box).

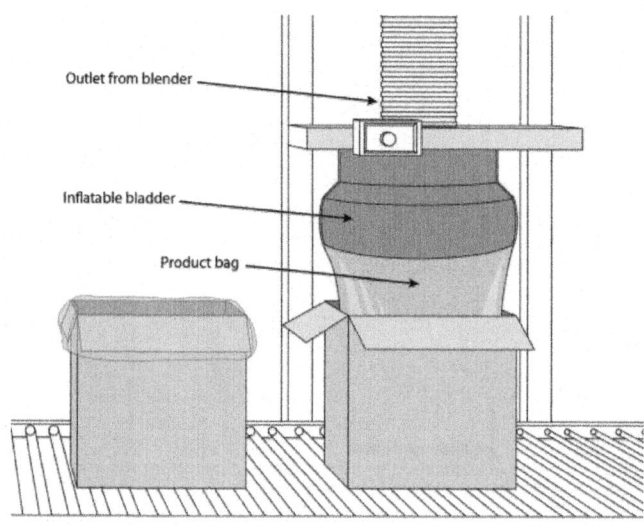

Outlet from blender

Inflatable bladder

Product bag

Figure 15: An inflatable seal used to contain nanopowders/dusts as they are discharged from a process such as spray drying

Product is discharged into the liner through a butterfly valve on the process outlet. Once full, the top of the first liner sleeve is closed using tape or a fastener, or it is heat sealed and cut. The product is sealed within the poly-lined container, and a new sealed poly liner is pulled down to start discharge into the next container. This continuous process seals off the primary leak paths for dust during unloading of an industrial blender or other equipment. These systems are commonly used in the pharmaceutical industry and may provide cost-effective alternatives to traditional local exhaust ventilation control systems for nanotechnology facilities.

Figure 16: A continuous liner product off-loading system that uses a continuous feed of bag liners fitted to the process outlet to isolate and contain process emissions and product (Used with permission from ILC Dover.)

Bag Dumping/Emptying

Technology used to control dusts during bag dumping has been in place for many years. The standard control—a ventilated bag dump station—consists of a hopper outfitted with an exhaust ventilation system to pull dusts away from workers as they open and dump bags of powdered materials. This equipment eradicates the dust problems caused by manually emptying bags and the need to dispose of empty bags. This ensures a healthy environment is maintained in the process area as well as reduces maintenance and repair problems caused by powder contamination to surrounding areas. The basic equipment consists of a bag dump cabinet with a dust extraction outlet for connection to a separate dust collector or existing plant exhaust. A bag is placed on the mesh support shelf and manually slit, with the contents falling directly into the inlet of a flexible conveyor or mixing tank. A side-mounted empty bag compactor may also be included. Design examples for these devices are available from several manufacturers of industrial materials. The British HSE has developed a control approach for a ventilated station for emptying bags of solid materials [HSE 2003d]. The control includes the specification of a face velocity of 200 fpm (1.0 m/s) and includes a waste bag collection chute (Figure 17).

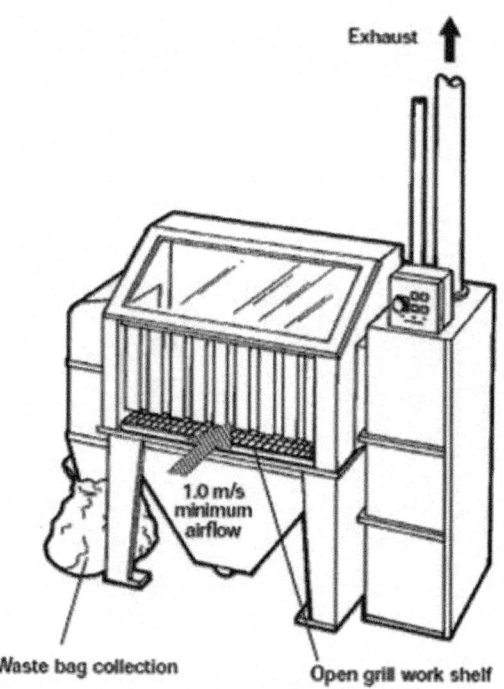

Figure 17: A ventilated bag-dumping station that reduces dust emissions when emptying product from bags into a process hopper (Contains public sector information published by the Health and Safety Executive and licensed under the Open Government License v1.0.)

Research into the effectiveness of these types of devices has shown that worker exposure to dust and vapors can be reduced. A review of commercially available units showed that their use with a variety of materials—including limestone, carbon black, and asbestos—controlled particle concentrations to acceptable levels [Heitbrink and McKinnery 1986]. However, particle contamination on the surface of the bag and handling/disposal of bags caused increased worker exposure. An integral pass through to a bag disposal chute/compactor can help reduce dust exposure resulting from bag handling. Further studies in mineral processing plants showed that the use of an overhead air supply also significantly decreased worker exposure [Cecala et al. 1988]. The ACGIH Industrial Ventilation Manual also has two designs that are applicable to the control of powder materials during bag dumping [ACGIH 2013]. Design plate VS-15-20, Toxic Material Bag Opening, is similar in design to the HSE station described above but recommends a slightly higher control velocity of 250 fpm at the face of the station opening. In addition, Design plate VS-50-10, Bin and Hopper Ventilation, requires a hood face velocity of 150 fpm. In general, higher velocities are specified to adequately capture dusts in a plant environment. While the materials used in the studies discussed above were not nanoscale, the application of the dust control concept is still relevant. However, the capture velocities specified in the Industrial Ventilation Manual may be excessive when attempting to contain nanomaterials; lower velocities may be warranted.

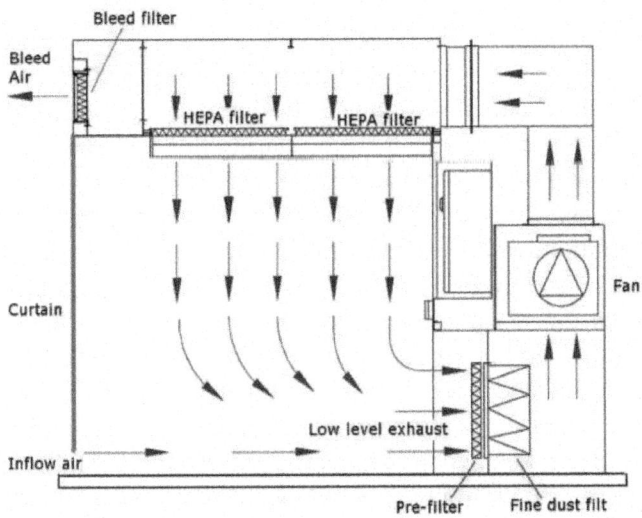

Figure 18: A unidirectional downflow booth for handling larger quantities of powders (Used with permission from Esco Technologies, Inc. [2012].)

Large-scale Material Handling/Packaging

Unidirectional flow booths, or downflow booths, as seen in Figure 18, are used in pharmaceutical applications for large-scale powder packing, process loading, and tray dryer loading [Hirst et al. 2002]. Similar applications have been proposed for handling hazardous dye powders. In general, these booths supply air from overhead (commonly at 100 fpm) over the full depth of the booth. Particles generated by processes carried out in the booths are captured and carried to the exhaust registers, which are located along the back wall of the booth. For the nanotechnology industry, these booths may provide a flexible solution for several common processes, including packaging of materials, transferring materials between process containers, or loading materials into containers for post processing. Floura and Kremer [2008] evaluated a downflow booth used for transferring 25 kg of lactose (a surrogate pharmaceutical material) from drum to drum inside a downdraft booth. Air samples were collected in the operator's breathing zone and around the perimeter of the process during the transfer operation. The operator scooped the lactose powder from the initial drum into the final product drum until it was nearly empty and then carefully inverted the bag to pour the remainder of the contents into the final container. With no active ventilation controls, the concentration within the operator breathing zone averaged 2,250 µg/m^3. When the ventilation inside the booth was turned on, the breathing zone concentration was substantially reduced to an average of 1.01 µg/m^3. Finally, the authors evaluated the downflow booth with a ventilated collar added.

The ventilated collar surrounded the interface between the drums and exhausted air at a rate of 425 ft3/min. During this test, the initial drum was inverted and the powder materials were emptied by gravity with the operator massaging the materials into the final product drum. The operator's breathing zone concentration averaged 0.03 µg/m3 during this process. This study showed that the use of a downflow booth significantly reduced operator exposure during powder transfer processes and that adding a second level of LEV, the ventilated collar, further reduced the exposure by two orders of magnitude. HSE Control Guidance Sheet 202, Laminar Flow Booth, presents a design for powderhandling processes called a horizontal- or cross-flow design [HSE 2003b]. The concept behind the design is similar to the downflow booth except that air enters the booth from the booth face. Air moves across the back of the worker toward the back of the booth. An issue with the cross-flow design is the secondary airflow patterns caused by the presence of the operator in the booth. Additionally, if purity or cleanliness of the product is important, sweeping of the air across the operator could be problematic. These patterns may cause turbulent dispersion of dust in the booth and result in higher operator exposure or potential leakage, compared to the downflow booth, but may provide a reasonable control for some processes.

Nanocomposite

Machining Initial studies have shown that machining some nanocomposite materials can result in the release of nanoscale particles to the work environment. Engineering controls when machining materials are available for most common processes. They range from ventilation of handheld tools using a high velocity-low volume (HVLV) system to the use of wet cutting techniques commonly adopted for silica control during construction activities. The use of standard dust controls such as those described by the HSE for woodworking as well as those identified in the ACGIH Industrial Ventilation Manual for machining processes provide a source of guidance that can be used to identify controls for machining processes. Bello et al. [2009] showed that the use of wet suppression techniques during sawing of nanocomposites reduced exposures down to background levels.

Maintenance

Tasks Maintenance of the production facility and equipment can lead to exposures that are often overlooked. Demou et al. [2008] noted that maintenance procedures were a source of considerable particle emissions, specifically during the vacuuming of a reactor using a vacuum cleaner with a high-efficiency filter. However, other researchers have observed that cleaning the process area after CNT preparation reduced airborne particle concentrations [Lee et al. 2010]. Another typical activity not reported in the literature is the changeout of facility air filters. When local exhaust ventilation is used to contain nanomaterials and dusts, facilities will typically use air filtration prior to exhausting air from the building or recirculating into the work zone. When filters require change-out, the use of integral containment equipment and procedures can reduce maintenance worker exposure. Other general maintenance procedures, such as modifying ductwork or performing fan maintenance, will also require appropriate precautions to avoid exposing workers to nanomaterials settled in the equipment. In addition, general good housekeeping processes and written spill response procedures can help reduce the potential for worker exposure.

Filter Change–out–BagIn/Bag Out Systems

Bag in/bag out procedures are typically designed to protect workers performing maintenance on air filter change out. Bag in/bag out housings are specifically designed to allow for removal of a dirty air filter while minimizing worker exposure [Filtration Group Inc. 2012]. In these systems, a plastic liner is attached to a service port on the filter unit, as shown on the following page in Figure 19. When the filter is ready for replacement, the facility maintenance worker, wearing appropriate PPE, removes the filter into a liner. This process contains the filter with its contaminants so the worker is not exposed and the particulates are not resuspended in the workplace environment.

Spill Cleanup Procedures

An organized, clean workplace enables faster and easier production, improves quality control, and reduces the potential for exposure. It is important to maintain good general housekeeping practices so that leaks, spills, and other process integrity problems are readily detected and corrected. Proper practices regarding spills include:

- Allowing only individuals wearing appropriate protective clothing and equipment and who are properly trained, equipped, and authorized for response to enter the affected area until the cleanup has been completed and the area properly ventilated.
- Using HEPA-filtered vacuums, wet sweeping, or a properly enclosed wet vacuum system for cleaning up dust that contains nanomaterials.
- Cleaning work areas regularly with HEPA-filtered vacuums or with wet sweeping methods to minimize the accumulation of dust.
- Cleaning up spills promptly.
- Limiting accumulations of liquid or solid materials on work surfaces, walls, and floors, to reduce contamination of products and the work environment.

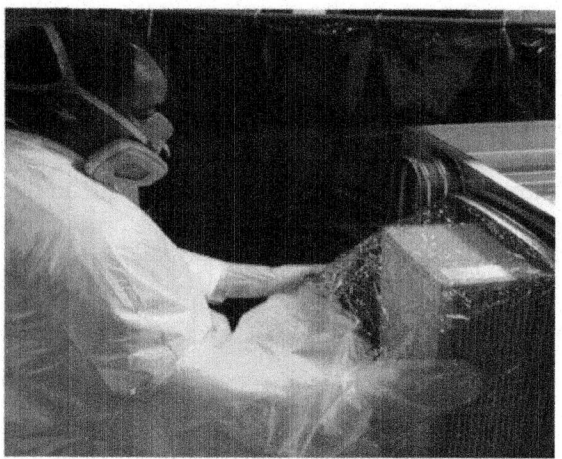

Figure 19: Removal of a dirty air filter from a ventilation unit into a plastic bag to minimize worker exposure to particles captured by the filter unit (Used with permission from Filtration Group Inc. [2012])

CONTROL EVALUATIONS

The effectiveness of engineering controls for reducing exposures to nanomaterials during manufacturing and handling has not been widely investigated. To evaluate control measures in nanomanufacturing facilities, investigators need to collect both quantitative and qualitative data to describe nanoparticle emissions. Accurate direct-reading instruments allow investigators to identify the source of contamination in real time for various task scenarios. More detailed information about the materials, such as morphology and chemical characteristics, can be obtained by collecting air filter samples for off-line analysis.

Approaches to Evaluation

Strategies for measuring nanomaterial exposures and emissions in the workplace are being developed and evaluated by a range of researchers [Brouwer et al. 2009; NIOSH 2009a; OECD 2009; Ramachandran et al. 2011]. Because there are currently no exposure limits for engineered nanomaterials in the United States, a multifaceted approach combining qualitative analysis with quantitative means should be used to determine nanoparticle emissions and control effectiveness [Oberdörster et al. 2005]. However, some researchers have suggested using non-mass-based metrics such as surface area or particle number as a reasonable approach to assessing health effects [Wittmaach 2007; Rushton et al. 2010; Oberdörster et al. 2005]. The evaluation procedures include (1) identification of emission sources, (2) background and area monitoring, (3) air concentration measurement by directreading instruments and filter-based sampling, and (4) measurement of air velocity and patterns.

Identification of Emission

Sources The main purpose of the initial assessment or walk-through survey is to identify potential sources of emissions and to help researchers prepare a sampling plan for the in-depth evaluation of processes and control measures. Portable direct-reading devices (e.g., handheld CPCs and photometers) are recommended for quick identification. The initial assessment should involve looking at the processes and equipment as well as the general plant environment. To optimize and improve engineering controls, a control checklist is recommended for collecting basic information on methods, manufacturing processes, and existing controls.

Background and Area Monitoring

A plan to assess control effectiveness requires that measurements are first taken of background concentrations in adjacent work areas. This allows the contribution of individual processes to be assessed by removing the background component [Brouwer et al. 2004; Demou et al. 2008; Peters et al. 2009]. The background measurement should be repeated after process or task evaluations. High background concentrations need to be addressed before control evaluation. The following factors can affect background data:

Monitoring period. In a nanomanufacturing facility, the day of the week or the time of day during the monitoring period will affect background levels since worker movement and frequency of worker operations are variable.

- Other activities or operations around the monitored activity. Any operation, such as product harvesting or equipment maintenance, in areas outside the monitoring location could potentially influence background concentrations at the monitoring location.

- General ventilation conditions. The layout and operation of the general ventilation system in the workplace should be considered while monitoring background concentrations. Basic ventilation data (e.g., volume of air flow, location of supply and exhaust, general air movement in the facility), including air supply source, should be collected. Additionally, variations in environmental conditions (especially humidity) need to be measured.

- Other sources. Some equipment can produce incidental (nonprocess) nanoparticles. Examples include diesel engines, welders, gas-fired heaters, and air compressors for pulse-jet baghouses or dust collectors.

Area (or static) monitoring can also be conducted to evaluate the general air quality of workplaces. Instruments such as the CPC and impactors are suitable for this type of monitoring. Ideally, filter samples can be taken at the same location as area monitoring with direct-reading instruments to make a side-by-side comparison.

Air Monitoring and Filter Sampling

The selection of direct-reading instruments (Table 3) for field evaluation must cover a wide range of particle sizes. Particle diffusion occurs rapidly when nanoparticles are released in the workplace. It results in nanomaterial agglomerates because of particle collisions. For example, the average particle size (or size distribution) is larger during product transfer than right after product harvesting. Based on the data collected during initial assessment, characterization of nanomaterial emissions can be conducted with direct-reading instruments to provide higher resolutions of spatial and time variation. To evaluate control efficiency for specific processes or tasks, the sampling ports should be located as close as possible to the suspected emission sources but outside of control measures (or at a worker's breathing zone). Filter sampling for off-line qualitative analysis must occur in parallel with real-time monitoring. Sampling duration may not be an issue for most direct-reading instruments but should be considered for filter sampling to avoid overloading. The data collected from an initial assessment can be used to determine sampling time and flow rate for filter samples.

Table 3: Summary of instruments and techniques for monitoring nanoparticle emissions in nanomanufacturing workplaces

Metric	Instrument	Remarks
Aerosol concentration	CPC	Real-time measurement. Typical concentration range of up to 400,000 particles/cm³ for stand-alone models with coincidence correction; 100,000 particles/cm³ for hand-held models.
	DMPS	SMPS often uses a radioactive source. FMPS uses electrometer-based sensors. Concentration range from 100–10⁷ particles/cm³ at 5.6 nm and 1–105 particles/cm³ at 560 nm.
Surface area	Diffusion charger	Need appropriate inlet pre-separator for nanoparticle measurement. Total active surface area concentration up to 1,000 μm²/cm³.
	ELPI	Real-time size-selective detection of active surface area concentration. 2×10⁴–6.9×10⁷ particles/cm³ depending on size range/stage.
Mass	Size selective static sampler	Low pressure cascade impactors. Micro-orifice impactors.
	TEOM	EPA standard reference equivalent method.
Aerosol concentration by calculation	ELPI	
Surface area by calculation	DMPS	
	DMPS and ELPI used in parallel	Surface area is estimated by difference in measured aerodynamic and mobility diameters.
Mass by calculation	ELPI	Calculated by assumed or known particle charge and density.
	DMPS	Calculated by assumed or known particle charge and density.

Abbreviations: CPC=condensation particle counter; DMPS=differential mobility particle sizer; SMPS=scanning mobility particle sizer; FMPS=fast mobility particle sizer; ELPI= electric low pressure impactor; TEOM=tapered element oscillating microbalance

Assessment of Air Velocities and Patterns

The measurement of air velocity and pattern is important to establish sampling locations, evaluate outdoor contaminant penetration, and assess the performance of existing control measures. Two widely used air fluid velocity measuring devices are the Pitot tube and the hot-wire anemometer. The Pitot tube is useful to measure flow in ducts with high temperatures and/or high particle concentrations, which could damage the thermal anemometer probe. Shown in Figure 20, a Pitot tube is a primary standard that measures total and static pressures, and air velocity is calculated by using the pressure difference (i.e., velocity pressure) based on the Bernoulli equation. The method for conducting a Pitot traverse is described in the ACGIH Industrial Ventilation Manual [ACGIH 2013]. The Pitot traverse is typically used to measure duct air velocity to estimate overall system exhaust flow rate. Occasionally it is difficult to find a suitable location for Pitot tube traverses. Accurate duct velocity can be obtained using this method; however, poor measuring locations will cause inaccurate estimates of exhaust air flow. Sometimes the airflow through a hood can only be determined by measuring the air velocity at the hood face. The measurement of fume hood face velocity is an important method to assess proper operation and containment. The average hood face velocity can be measured by dividing the opening of the hood into equal area grids of approximately one square foot and logging the velocity at the center of each grid with the thermal anemometer. To measure the velocities at each grid point, the anemometer should be held perpendicular to the direction of air flow. An average face velocity can be calculated while the variation in hood velocity from grid to grid should be assessed and noted [ACGIH 2013; ASHRAE 1995].

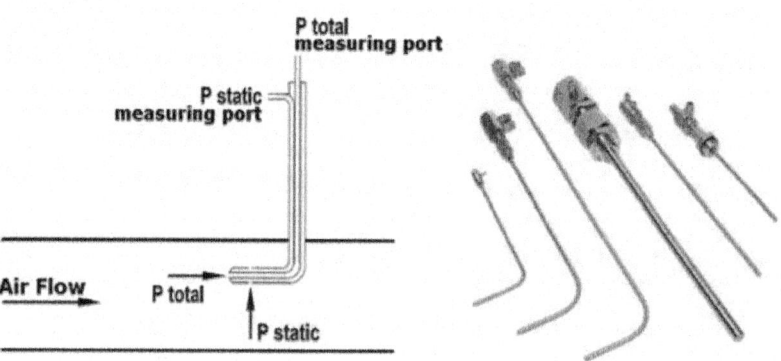

Figure 20: Operating principle of a Pitot tube (left) and different types of Pitot tubes (right)

In addition to Pitot tubes and anemometers for measuring air velocity, smoke generators provide a low-cost method to visualize airflow patterns around control measures. Figure 21 shows an example of a smoke generator. Airflow visualization techniques can be used to help understand the patterns of airflow in and around exhaust hoods and pressure differences between adjacent areas/rooms. Smoke can be released around the edge of, or inside of, a local exhaust hood to visualize the airflow patterns.

This will help determine whether airborne particles are being effectively captured and removed by the ventilation system. Recorded observations should concentrate on (1) how much of the smoke is entrained into the LEV, (2) how quickly the exhaust captures the smoke, (3) the direction of air flow, and (4) whether or not any of the smoke visibly enters the worker's breathing zone. In addition, multiple replications of smoke-release observations should be made at locations where LEV performance is marginal or poor as indicated by reverse airflow, lack of air movement, slow clearance time, and escape of smoke from the hood.

Special attention should be paid in subsequent tracer gas testing and air velocity measurements to locations where smoke release observations indicate poor or marginal capture efficiency. In addition, video may be taken of airflow visualization tests to provide feedback information to the company on system performance and factors that negatively affect hood performance. Another use for airflow visualization is the evaluation of room pressurization status. It is recommended that rooms where nanomaterials are used be kept at an atmospheric pressure that is lower than adjacent areas.

This condition helps contain the materials and reduce exposures to workers in other areas of the plant. Smoke should be released at the interfaces (doors or other openings) between any nanomaterial production areas and attached spaces. By releasing smoke at these interfaces, it can be easily observed whether air is moving into or out of the production area and proper remediation approaches may be implemented where necessary.

To qualitatively assess whether exhaust re-entrainment may be an issue, smoke can be released within each hood in the production room while a researcher observes the emission of the smoke through the exhaust stack. This qualitative test will help to evaluate the potential for re-entrainment of exhaust into any air intakes or roof openings.

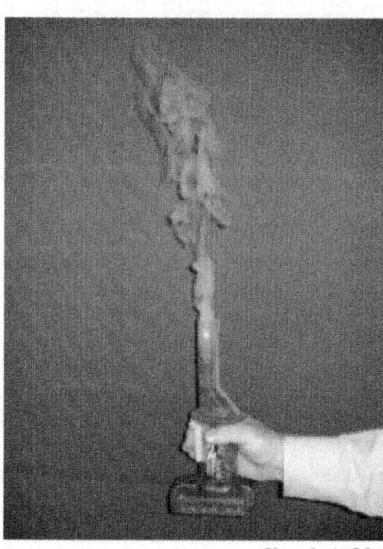

Figure 21: Smoke generator to visualize airflow

The behavior of the exhaust plume is dependent on varying environmental conditions such as wind speed and direction; therefore, this test should be repeated to capture the potential for re-entrainment under a variety of conditions. In addition, air velocity measurements should be taken at the center of the exhaust duct opening to evaluate the discharge velocity of the hood exhaust. These readings should be evaluated, along with the physical design and installation of the exhaust stack, against guidance from consensus standards organizations such as ASHRAE, ACGIH, or AIHA.

Facility Sampling and Evaluation

Checklist When evaluating a facility that manufactures or uses nanomaterials, it is important to first assess what engineering controls are in place in the facility. The initial assessment should involve looking at the processes and equipment as well as the general plant environment, the effective use of the engineering control by the operator(s), and the overall performance of the control equipment. Checklists are useful tools for helping to identify the process and facility factors related to nanomaterial production, use, emissions, and exposure. A checklist as shown in Table 4 may help for collecting basic process information (e.g., capacity, location, and usage) and control operation and maintenance parameters to ensure effectiveness of exposure control.

Table 4: Checklist of controls for nanomaterial manufacturing and handling

Item	Category	Data
Process/task	(select all applicable) ☐ Weighing ☐ Mixing ☐ Transferring ☐ Drying ☐ Cleaning ☐ Cutting/sanding ☐ Harvesting ☐ Unpacking ENMs ☐ Maintenance/repair ☐ Finishing (drilling, sawing, grinding) ☐ Packaging/shipping ☐ Others: _____	Workspace Duration (min) Frequency (times per day) Number of workers involved PPE type
	Background	Concentration Size distribution: _____ Number: _____ Mass: _____
Nanomaterial	☐ SWCNT ☐ MWCNT ☐ Other carbon-based ☐ Metals ☐ Oxides ☐ Quantum dots ☐ Composite: _____ ☐ Others: _____	Processing rate/volume Primary particle size Concentration at source Number: _____ Mass: _____ Concentration at worker breathing zone or area (designate) Number: _____ Mass: _____ Breathing zone: _____ Area: _____

Item	Category	Data
Control type	☐ NONE ☐ Local exhaust ☐ General exhaust/dilution ☐ Ventilated enclosure ☐ Fume hood ☐ Dust collector ☐ Laminar room ☐ Glove box ☐ Booth ☐ Other:_____	Dimensions Location Operation ☐ Hood type ☐ Face velocity: _____ ☐ Flow rate: _____ ☐ Temp: _____ ☐ Enclosure integrity ☐ Airflow patterns ☐ Recirculation Fan/filtration information Filter type: _____ Manufacturer: _____ Resistance (pressure drop): _____ Nominal design flow rate: _____ Fan type: _____ Flow rate: _____ Stack position/design:
Visual observation	Workspace Surface contamination Housekeeping Layout	

Industrial Exhaust Ventilation Deficiency Report Worksheet

☐ Building:	☐ Room:	☐ Hood number:
☐ Date:	☐ Investigator/reporter:	☐ Fan number:

Notes and sketch

Management

- ☐ No local cognizant person
- ☐ Lack of records
- ☐ Lack of up-to-date plans and specifications
- ☐ Lack of emergency plan
- ☐ Insufficient employee training
- ☐ No hood testing mechanism
- ☐ No hood-use approval mechanism

Ductwork

- ☐ Holes, air leaking
- ☐ Dents
- ☐ Poor construction
- ☐ Plugged
- ☐ Corroded
- ☐ Leaking
- ☐ Dampers improperly set
- ☐ Fire dampers
- ☐ Doesn't meet SMACNA qualifications

Hood

- ☐ Improper type for operation/chemicals used
- ☐ Air leaking from hood (smoke noncontainment)
- ☐ Surfaces corroded
- ☐ Surfaces dirty
- ☐ Hood mechanisms inoperable
- ☐ Lack of real-time airflow monitor
- ☐ Flammable construction materials
- ☐ Slots not open to appropriate size
- ☐ Slots blocked by equipment, chemicals

Fan/Motor

- ☐ Worn out or corroded
- ☐ Insufficient rpm
- ☐ Belts slipping or broken
- ☐ Motor burned out
- ☐ Undersized fan

Hood operations

- ☐ Use of hood when hood exhaust off
- ☐ Hood not being used
- ☐ Inappropriate materials/equipment in hood
- ☐ Noisy

Stack

- ☐ Not attached
- ☐ Inappropriate location
- ☐ Inadequate height
- ☐ Stack exit velocity insufficient
- ☐ Aesthetic enclosure hinders dispersion

Work practices

- ☐ Untrained personnel
- ☐ Rapid movements at hood face
- ☐ Placing upper body in hood
- ☐ Operating outside hood

Exhaust hood

- ☐ Inadequate exhaust volume
- ☐ Inadequate face velocity
- ☐ Inadequate face velocity range
- ☐ Turbulence in hood face

Industrial Exhaust Ventilation Deficiency Report Worksheet (Continued)

Make-up air	System maintenance
☐ No replacement air ☐ Insufficient air for dilution of fugitive emissions ☐ Contaminated by exhaust air ☐ Supply diffuser blows on hood face ☐ Supply diffuser blocked ☐ Temperature inadequate ☐ Employee complaints (noise, draft) ☐ Does not meet ASHRAE 62 provisions ☐ Supply not balanced with exhaust	☐ Inadequate maintenance (equipment broken) ☐ Lack of ongoing PM program
Worksite	**Manifold exhaust systems**
☐ Cluttered, housekeeping poor, dirty ☐ Hood positioned near door, window, walkway, other turbulence ☐ Fire escape routes blocked ☐ Aisles blocked	☐ Likelihood of fire/explosion; mixed chemicals ☐ Corrosion in manifold ☐ Condensation in manifold ☐ One hood goes positive ☐ Part of system under positive pressure

EVALUATING SOURCES OF EMISSIONS AND EXPOSURES TO NANOMATERIALS

Direct-reading Monitoring

Currently, it is unclear which metrics associated with exposures to engineered nanomaterials are most important from a health and safety perspective. The mass-based metric is traditionally used to characterize toxicological effects of exposure to air contaminants. Animal in vivo exposure studies and cell-culture-based in vitro experiments show that size and shape are the two major factors influencing toxicological effects of engineered nanomaterials. Some of the instruments developed to characterize nanoparticles are capable of real-time measurement [Brouwer et al. 2004; Pui 1996; Ramachandran 2005]. Real-time measurement of aerosolized particles, including primary nanoparticles and agglomerates, play an important role in identifying nanomaterial emissions and evaluating control systems during field surveys. The measuring devices used to evaluate controls in the workplace should be portable and robust. Information about readily available instruments and techniques for nanoparticle monitoring (Table 3) has been summarized and discussed in technical reports [BSI 2007a; EU-OSHA 2009; HSE 2004; ISO 2007, 2008; Mark 2007; Park et al. 2010a, b, 2011]. It is noted that some of the instruments on the list in Table 3 are not suited for monitoring nanomaterial emissions in the workplace. For instance, the tapered element oscillating microbalance (TEOM) is used by the Environmental Protection Agency as a standard reference equivalent method to monitor environmental air quality, but the cut-off particle sizes of

10, 2.5, or 1 µm and dimensions of this instrument limit its use for workplace sampling. Another example is the scanning monitoring particle sizer (SMPS), which uses a radioactive source to bring the sampling aerosol to charge equilibrium. This can make shipping difficult. Sometimes it can be difficult to obtain quantifiable mass concentrations of nanomaterials in the workplace using impactor sampling. Newly developed devices, such as photometers, can detect nanoparticles as small as 50−100 nm with resolution around 1 µg/m3 . These instruments can provide continuous monitoring for real-time mass concentrations. Data from direct-reading instruments only provide a semiquantitative indication of potential nanoparticle emissions. Fluctuating background concentrations may make determination of control efficiency difficult; changes in background concentration may lead the evaluator to think that the controls are performing either better or worse than they are actually performing. In addition, directreading instruments cannot distinguish particle source and composition; these can only be determined through off-line microscopic and chemical analysis. Sampling quality is always an issue for field evaluation. High-quality sampling results can be obtained by following certain steps. The sampling data can only be trusted by using instruments that have been calibrated for nanoparticle sampling before use. Factory calibration for particle counters and sizers typically uses reference materials having a range of particle sizes. If possible, the instruments should be calibrated with the target nanomaterials in the laboratory before using them for field study. The comparison calibration should also be done on identical instruments if they will be used in a field survey. To maintain consistent sampling performance, a zero check for instruments should be performed before daily use and after sampling high-particle emissions. Sampling loss due to particles deposited in sampling tubes can be lowered by using conductive tubing and minimizing tubing length and bends in the tubing. The sampling location should be considered carefully, because nanoparticles diffuse rapidly through the workplace air. The choice of sampling location could have a large influence on the sampling results. The sampling ports must be kept as close as possible to the emission source.

Off-line Analysis

In addition to direct-reading instrument measurements, nanoparticle emissions can also be characterized using off-line analysis techniques. Off-line analysis methods can determine the physical and chemical properties of airborne nanomaterials, such as particle size, shape, surface area, composition, and agglomeration state. These properties are useful to evaluate exposure and toxicology of nanomaterials in the workplace. Off-line analysis can also be useful in separating background nanomaterials from engineered nanomaterials, based on size, shape, morphology, etc. NIOSH has developed techniques for off-line analysis using filter samples. NIOSH Method 7402 (Asbestos by TEM) was developed to collect filter samples of materials with large aspect ratios for analysis using transmission electron microscopy (TEM) and can be used to determine particle morphology and geometry. NIOSH Method 5040 (Diesel particulate matter as elemental carbon) can be used to measure elemental carbon (e.g., CNT, CNF). Other nanomaterials (e.g., metals) can be collected on filters and analyzed using NIOSH Method 7300 (Elements by ICP). Using the mass determined by chemical analysis and dividing by the total air flow volume will provide a mass concentration of the nanomaterial of interest. As with real-time instrumentation, background samples are collected to help distinguish nanomaterials from incidental ultrafine aerosols. The optical diameters of single particles and agglomerates can be compared to data from direct-reading instruments discussed above. Filters overloaded with particles cannot be analyzed by direct-transfer TEM analysis. Therefore, filter sample volume needs to be balanced against the particle emission rate to avoid filter overload. The results of the initial walk-through survey with portable particle counters should provide basic information to help determine appropriate filter sampling volume and collection time.

Video Exposure Monitoring

Video exposure monitoring (VEM) is an exposure assessment technique in which real-time monitoring devices (e.g., nanoparticle and dust monitors) are synchronized with video of the work activity [Beurskens-Comuth et al. 2011]. The product of VEM is a video of the work activity with a graphical presentation of exposure concentrations that corresponds to the job task displayed on the video. VEM aides in the identification of work practices that can contribute significantly to overall exposure patterns by giving a visual display of work activities and the corresponding real-time monitoring values. With this exposure assessment tool, both management and employees can be shown which activities have the highest exposure concentrations and can therefore benefit from a change in work practice, installation of engineering controls to mitigate the exposure, or the use of PPE. The VEM method was initially developed by NIOSH engineers in the late 1980s to bring together work activity data (video recordings) with direct-reading exposure data. By identifying the critical activities that contribute most to a worker's exposure, sampling resources can be directed to controlling those job activities that affect exposures. Work activity variables can also be keyed into the exposure database to statistically assess the impact of work activities on exposures. The method permits researchers and safety and health professionals to capitalize on the time element of the direct-reading data by uniting the exposure measurement with the corresponding work activities. The VEM method allows directreading monitors to be used as more than simple detectors and have a significant impact on occupational exposures.

Evaluating Ventilation Control Systems

Several methodologies are available to evaluate local exhaust systems and other exposure controls. These techniques include indirect approaches, such as the measurement of capture velocity, slot velocities, hood static pressure, and other system performance parameters [Goodfellow and Tahti 2001]. Often these measures are compared with design guidance or standards from organizations such as ASHRAE, ANSI, AIHA, and ACGIH. In general, these tests provide a method of checking system performance without the requirement for expensive instrumentation or a high level of operator experience. Because these measures do not directly assess system performance, it is often a good idea to use methods that are more specialized than these indirect methods. One method commonly used to evaluate the capture efficiency of the LEV system is the quantitative capture test. Tracer gas release and measurement is a method used to quantitatively estimate the efficiency of industrial exhaust ventilation hoods [Hampl 1984; Hampl et al. 1986; Marzal et al. 2003b]. This method typically involves using a surrogate for the process-generated contaminant and requires the use of special measurement and dispersion equipment to conduct the test. A variety of tracers have been used, including oil mist aerosols, polystyrene latex spheres, and gases [Beamer et al. 2004; Ellenbecker et al. 1983; Hampl 1984]. In addition to the quantitative capture method, qualitative methods, such as smoke release or dry ice tests, are often used to evaluate air movement. Smoke generation and capture is a method often used to qualitatively evaluate the performance of ventilation controls [Marzal et al. 2003a; Woods and Mckarns 1995]. With this method, a source is used to introduce smoke in and around the hood. This allows the researcher to better understand the performance of the hood and evaluate the effect of cross currents on the capture of contaminants. These tests not only give the experimenter a sense of the system performance but provide invaluable information on where other measurements, such as air velocities and tracer gas experiments, should be concentrated. This testing is often conducted while workers are not in the production area, either after the work shift or while workers are on break.

Standard Containment Test Methods for Ventilated Enclosures

Some standard test methods (Table 5) to evaluate fume hoods have been developed: Invent-UK method, DIN 12924, BS 7258, EN 14175:2003, and

ANSI/ASHRAE 110-1995. One major difference between ANSI/ASHRAE 110-1995 and other standard test methods is that only one sampling probe is used to detect the test gas concentration near the worker's breathing zone. Other test methods adopt multiple sampling probes connected to a manifold to obtain the area concentration near the fume hood opening. The test methods of DIN 12924 and ANSI/ASHRAE 110-1995 use a manikin to test the containment effectiveness of fume hoods. Dynamic test conditions are specified in the test methods of EN 14175:2003 and ANSI/ASHRAE 110-1995. The purpose of the dynamic test is to evaluate the hood during typical maneuvers such as raising or lowering the sash and simulating the airflow disturbance related to a person walking in front of the hood. During field evaluations, ventilated enclosures should also be tested during normal-use conditions. Collecting samples both inside and outside the containment opening and in the worker's breathing zone is recommended to assess control effectiveness when workers are performing standard tasks.

Table 5: Comparison of the fume hood performance test methods

Test method	Invent-UK	DIN 12924	BS 7258	EN 14175:2003	ANSI/ ASHRAE 110-1995
Country	United Kingdom	Germany	Great Britain	European Union	United States
Test parameters	Face velocity	Tracer gas test	Tracer gas test	Face velocity	Face velocity and cross draft
	Tracer gas test			Tracer gas test	Smoke visualization
				Robustness test (dynamic test by walk-bys and traffic)	Tracer gas test
Tracer gas	10% SF_6 *+ 90% N_2 @ 3.0 LPM	10% SF_6 + 90% N_2 @ 3.33 LPM	10% SF_6 + 90% N_2 @ 2.0 LPM	10% SF_6 + 90% N_2 @ 2–4 LPM	100% SF_6 @ 4.0 LPM
Tracer gas sampling probes	9	20	Multi-probes depending on opening size	Multi-probes (inner and outer grids)	1 in breathing zone
Use of manikin	No	Yes	No	No	Yes

CONCLUSIONS AND RECOMMENDATIONS

Engineered nanomaterials are materials that are intentionally produced and have at least one primary dimension less than 100 nanometers (nm). Nanomaterials have properties different from those of the bulk material, making them unique and desirable for specific processes. These same properties may also cause adverse health effects in workers. Currently, the toxicity of many nanomaterials is unknown, but initial research indicates that there may be health concerns related to occupational exposures. Due to the potential for health effects, it is important to control worker exposures to the extent possible. The following are conclusions and recommendations for reducing the potential for employee exposures during nanomanufacturing processes based on current knowledge.

General

Hazards involved in processing and manufacturing nanomaterials should be managed as part of a comprehensive occupational safety and health management plan. Preliminary hazard assessments (PHAs) should be conducted to determine the need for control measures during the planning stage. Hazard assessments should be done during the operation of a facility and regularly updated when any processes change.

The concept of Prevention through Design (PtD) is to design out or minimize hazards early in the design process. When PtD is implemented, the control hierarchy is applied by designing safety into the work environment to prevent work-related injuries and illnesses.

Control Banding

With the absence of OELs, control banding is a potentially useful concept in the risk management of nanomaterials. Control banding is not intended to be a substitute for OELs and does not alleviate the need for environmental monitoring or industrial hygiene expertise.

Hierarchy of Controls

The hierarchy of controls should be followed when controlling potential occupational hazards from nanoparticles. Elimination and substitution are at the top of the hierarchy. However, eliminating nanomaterials may not be possible as the nanomaterials were likely chosen because of their unique properties. The manner in which these materials are handled and processed can largely affect the overall safety of the process The substitution of less hazardous materials for those that are a higher hazard should be considered to reduce the risk to workers. Substitution also applies to the form of the product used; for example, a slurry with less exposure potential could be used to replace a dry powder.

Engineering Controls

If elimination and substitution are not feasible to reduce hazards, engineering controls should be implemented. These could include local exhaust ventilation, isolation measures, and application of water or other material for dust suppression.

Engineering controls are likely the most effective control strategy for nanomaterials. Common controls used in the nanotechnology industry include fume hoods, biological safety cabinets, glove box isolators, glove bags, bag dump stations, and directional laminar flow booths. Each of these controls should be carefully designed and operated properly to be effective.

Preventative maintenance schedules should be developed to ensure that engineering controls are operating at design conditions.

Non-ventilation engineering controls cover a range of controls (e.g., guards and barricades, material treatment, or additives). These controls should be used in conjunction with ventilation measures to provide an enhanced level of protection for workers. Many devices developed for the pharmaceutical industry, including isolation containment systems, may be suitable for the nanotechnology industry. a. The continuous liner system allows filling product containers while enclosing the material in a polypropylene bag. This system should be considered for off-loading materials when the powders are to be packed into drums. b. Water sprays may reduce respirable dust concentrations generated from processes such as machining (e.g., cutting, grinding). Machines and tooling, as well as the material being cut or formed, must be compatible with water. If a fluid other than water is used, attention should be given to the fluid being applied to avoid creating a health hazard to workers.

A variety of controls are currently commercially available for use.

A checklist that collects basic process information (e.g., capacity, location, and usage) and control operation and maintenance parameters can optimize and improve existing exposure control. An example checklist is provided in Table 4.

Administrative Controls

Administrative controls and PPE are frequently used with existing processes where hazards cannot be effectively controlled solely with engineering controls. This could occur when control measures are not feasible or do not reduce exposures to an acceptable level. Administrative controls and PPE programs may be less expensive to establish but, over the long term, can be very costly to sustain. These methods for protecting workers have proven to be less effective than other measures and require significant efforts by the affected workers. A program that addresses the hazards present, employee training, and PPE selection, use, and maintenance should be in place when PPE is used. Administrative controls and PPE can also be useful for redundancy, especially in high-hazard situations. While engineering controls serve as primary controls, the administrative and PPE controls provide backup. Employers should implement the following work practices to control worker exposure to nanomaterials:

Educate workers on the safe handling of engineered nanomaterials to minimize the likelihood of inhalation exposure and skin contact.

Provide information to workers on the hazardous properties of the nanomaterials being produced or handled with instruction on how to prevent exposure.

Obtain the material safety data sheets (MSDS) when using nanomaterials from an outside source and review the information with employees who may come in contact with the materials. Given the lack of complete health information of many nanomaterials, the MSDS may not provide adequate guidance and should be assessed by the health and safety office.

To reduce the potential for release of nanomaterials, consider transferring powdered materials to a slurry, where possible.

Clean up spills of nanomaterials immediately and in accordance with written procedures. Appropriate PPE should be donned while performing clean-up tasks.

Provide additional control measures (e.g., a buffer area, decontamination facilities located by the hazard) to ensure that engineered nanomaterials are not transported outside the work area. Place a sticky mat at the exits of production areas to reduce the likelihood of spreading nanomaterials.

Encourage workers to use hand-washing facilities before eating, smoking, or leaving the worksite.

Provide facilities for showering and changing clothes to prevent the inadvertent contamination of other areas (including take-home) caused by the transfer of nanomaterials on clothing and skin.

Prohibit the consumption of food or beverages in work areas where nanomaterials are handled.

Ensure work areas and equipment, e.g., balance, are cleaned at the end of each work shift, at a minimum, using either a HEPA-filtered vacuum cleaner or wet wiping methods. Dry sweeping or compressed air should not be used to clean work areas. Cleanup should be conducted in a manner that prevents worker contact with wastes. Disposal of all waste material should comply with all applicable federal, state, and local regulations.

Store nanomaterials, whether suspended in liquids or in a dry particle form, in closed (tightly sealed) containers whenever possible.

Conduct routine industrial hygiene and medical monitoring to ensure that work practices and engineering controls are effective.

Personal Protective Equipment

Because nanoparticles have been found to penetrate the skin, items such as gloves, gauntlets, and laboratory clothing or coats should be worn when working with nanoparticles. Good hygiene practices for wearing the protective equipment should be followed.

Gloves made of neoprene, nitrile, or other chemical-resistant gloves should be used and changed frequently or whenever they are visibly worn, torn, or contaminated.

Respiratory protection should be used to reduce worker exposures to acceptable levels in the absence of effective engineering controls, during the installation or maintenance of engineering controls, for short-duration tasks that make engineering controls impractical, and during emergencies.

Respirators in the workplace should be used as part of a comprehensive respiratory protection program. The program should include written standard operating procedures; workplace monitoring; hazard-based selection; fit-testing and training of the user; procedures for cleaning, disinfection, maintenance, and storage of reusable respirators; respirator inspection and program evaluation; medical qualification of the user; and the use of NIOSH-certified respirators.

REFERENCES

1.   60 Fed. Reg. 30336 [1995]. National Institute for Occupational Safety and Health:Respiratory protective devices; final rule. (To be codified as 42 CFR Part 84.)

2.   63 Fed. Reg. 1152 [1998]. Occupational Safety and Health Administration: respiratory protection; final rule. (To be codified at 29 CFR 1910 and 1926.)

3.   ACGIH [2010]. Industrial ventilation: a manual of recommended practice for operation and maintenance. Cincinnati, Ohio: American Conference of Governmental Industrial Hygienists.

4.   ACGIH [2013]. Industrial ventilation: a manual of recommended practice for design.

5.   Cincinnati, Ohio: American Conference of Governmental Industrial Hygienists.

6.   Ahn K, Woskie S, DiBerardinis L, Ellenbecker M [2008]. A review of published quantitative experimental studies on factors affecting laboratory fume hood performance. J Occup Environ Hyg 5(11):735–753.

7.   ANSI/AIHA [2012]. Occupational health and safety management systems. Fairfax, VA: American Industrial Hygiene Association Publication No. ANSI Z10–2012.

8.   ASHRAE [1995]. Method of testing performance of laboratory fume hoods. Atlanta, GA: American Society of Heating Refrigerating and Air-Conditioning Engineers, Publication No. ANSI/ASHRAE 110 1995.

9.   ASHRAE [2011]. ASHRAE handbook—HVAC applications. Atlanta, GA: American Society of Heating, Refrigerating, and Air-conditioning Engineers.

10.   Bałazy A, Podgórski A, Gradoń L [2004]. Filtration of nanosized aerosol particles in fibrous filters. I–experimental results. J Aerosol Sci 35:967–980.

11.   Balazy A, Toivola M, Reponen T, Podgorski A, Zimmer A, Grinshpun SA [2006]. Manikinbased performance evaluation of N95 filtering-facepiece respirators challenged with nanoparticles. Ann Occup Hyg 50(3):259–269.

12.   Bayer MaterialScience [2010]. Occupational exposure limit (OEL) for Baytubes defined by Bayer MaterialScience. Leverkusen, Germany: Bayer MaterialScience.

13.   Beamer BR, Topmiller JL, Crouch KG [2004]. Development of evaluation procedures for local exhaust ventilation for United States postal service mail-processing equipment. J Occup Environ Hyg 1(7):423–429.

14.   Bello D, Wardle BL, Yamamoto M, Guzman de Villoria R, Garcia EJ, Hart AJ, Ahn K, Ellenbecker MJ, Hallock M [2009]. Exposure to nanoscale particles and fibers during machining of hybrid advanced composites containing carbon nanotubes. J Nanopart Res 11(1):231–249.

15.   Beurskens-Comuth PAWV, Verhist K, Brouwer D [2011]. Video exposure monitoring as part of a strategy to assess exposure to nanoparticles. Ann Occup Hyg 55(8):937–945.

16.   Brock B [2009]. Knowledge brief: containment hierarchy of controls. Tampa, FL: International Society of Pharmaceutical Engineers.

17.   Brouwer D [2010]. Exposure to manufactured nanoparticles in different workplaces. Toxicology 269(2):120–127.

18. Brouwer D, van Duuren-Stuurman B, Berges M, Jankowska E, Bard D, Mark D [2009]. From workplace air measurement results toward estimates of exposure? Development of a strategy to assess exposure to manufactured nano-objects. J Nanopart Res 11:1867–1881.

19. Brouwer DK, Gijsbers JHJ, Lurvink MWM [2004]. Personal exposure to ultrafine particles in the workplace: exploring sampling techniques and strategies. Ann Occup Hyg 48(5):439–453.

20. BSI [2007a]. Nanotechnologies, part 1: good practice guide for specifying

21. manufactured nanomaterials. Reston, VA: British Standards Institution, Publication No. PD 6699-1:2007.

22. BSI [2007b]. Nanotechnologies, part 2: guide to safe handling and disposal of manufactured nanomaterials. Reston, VA: British Standards Institution Publication No. PD 6699-2:2007.

23. BSI [2007c]. Occupational health and safety management systems; requirements.

24. Reston, VA: British Standards Institution, Publication No. BS OHSAS 18001:2007.

25. Cecala AB, Volkwein JC, Daniel JH [1988]. Reducing bag operator's dust exposure in mineral processing plants. Appl Ind Hyg 3(1):23–27.

26. Cena LG, Peters TM [2011]. Characterization and control of airborne particles emitted during production of epoxy/carbon nanotube nanocomposites. J Occup Environ Hyg 8(2):86–92.

27. CFR. Code of Federal Regulations. Washington, DC: U.S. Government Printing Office, Office of the Federal Register.

28. Chou CC, Hsiao HY, Hong QS, Chen CH, Peng YW, Chen HW, Yang PC [2008].

29. Single-walled carbon nanotubes can induce pulmonary injury in mouse model. Nano Lett 8(2):437–445.

30. Conti JA, Killpack K, Gerritzen G, Huang L, Mircheva M, Delmas M, Hathorn BH, Appelbaum RP, Holden PA [2008]. Health and safety practices in the nanomaterials workplace: results from an international survey. Environ Sci Technol 42(9):3155–3162.

31. Dahm MM, Yencken MS, Schubauer-Berigan MK [2011]. Exposure control strategies in the carbonaceous nanomaterial industry. J Occup Environ Med 53(6 Suppl):S68–73.

32. Davies CN [1977]. Aerosol science. London: Academic Press, p. 468.

33. Demou E, Peter P, Hellweg S [2008]. Exposure to manufactured nanostructured particles in an industrial pilot plant. Ann Occup Hyg

52(8):695–706.

34. DHHS [2009]. Biosafety in microbiological and biomedical laboratories (BMBL) 5th Edition. Cincinnati: U.S. Department of Health and Human Services, Public Health Service, Centers for Disease Control and Prevention, National Institutes of Health, Publication No. DHHS (CDC) 21–1112.

35. DiNardi SR [2003]. The Occupational environment: its evaluation, control, and management. Fairfax, VA: AIHA Press.

36. Elder A, Gelein R, Silva V, Feikert T, Opanashuk L, Carter J, Potter R, Maynard A, Ito Y, Finkelstein J, Oberdorster G [2006]. Translocation of inhaled ultrafine manganese oxide particles to the central nervous system. Environ Health Perspect 114(8):1172–1178.

37. Ellenbecker MJ, Gempel RF, Burgess WA [1983]. Capture efficiency of local exhaust ventilation systems. Am Ind Hyg Assoc J 44(10):752–755.

38. Eninger RM, Honda T, Reponen T, McKay R, Grinshpun SA [2008]. What does respirator certification tell us about filtration of ultrafine particles? J Occup Environ Hyg 5(5):286–295.

39. Esco Technologies Inc. [2012]. Pharmacon downflow booth [http://escoglobal.com/products/ download/1334055030.pdf]. Date accessed: November 11, 2012.

40. EU-OSHA [2009]. Literature review: workplace exposure to nanoparticles. Bilboa, Spain: European Agency for Safety and Health at Work, p. 89.

41. Evans DE, Ku BK, Birch ME, Dunn KH [2010]. Aerosol monitoring during carbon nanofiber production: mobile direct-reading sampling. Ann Occup Hyg 54(5):514–531.

42. Filtration Group Inc. [2012]. HEPA seal bag in/bag out operation and maintenance manual [www.filtrationgroup.com/../HEPA_BagIn_BagOut_Housing_Install_01.pdf]. Date

43. accessed: November 14, 2012.

44. Floura H, Kremer J [2008]. Performance verification of a downflow booth via surrogate

45. testing. Pharmaceut Eng 28(6):1–9.

46. Gao P, Jaques PA, Hsiao TC, Shepherd A, Eimer BC, Yang M, Miller A, Gupta B, ShafferR [2011]. Evaluation of nano- and submicron particle penetration through ten nonwoven fabrics using a wind-driven approach. J Occup Environ Hyg 8(1):13–22.

47. Genaidy A, Tolaymat T, Sequeira R, Rinder M, Dionysiou D [2009].

Health effects of exposure to carbon nanofibers: systematic review, critical appraisal, meta analysis and research to practice perspectives. Sci Total Environ 407(12):3686–3701.

48. Goodfellow H, Tahti E [2001]. Industrial ventilation design guidebook. San Diego, CA: Academic Press, p. 1519.

49. Hampl V [1984]. Evaluation of industrial local exhaust hood efficiency by a tracer gas technique. Am Ind Hyg Assoc J 45(7):485–490.

50. Hampl V, Niemela R, Shulman S, Bartley DL [1986]. Use of tracer gas technique for industrial exhaust hood efficiency evaluation—where to sample. Am Ind Hyg Assoc J

51. 47(5):281–287.

52. Heim M, Mullins BJ, Wild M, Meyer J, Kasper G [2005]. Filtration efficiency of aerosol particles below 20 nanometers. Aerosol Sci Technol 39:782–789.

53. Heitbrink WA, McKinnery WN Jr. [1986]. Dust control during bag opening, emptying and disposal. Appl Ind Hyg 1(2):101–109.

54. Hinds WC [1999]. Aerosol technology: properties, behavior, and measurement of airborne particles. New York: Wiley, p. 483.

55. Hirst N, Brocklebank M, Ryder M [2002]. Containment systems: a design guide. Woburn, MA: Gulf Professional Publishing, p. 199.

56. HSE [2003a]. Control guidance sheet 301: glovebox. In: COSHH essentials: easy steps to control chemicals. London: Health and Safety Executive.

57. HSE [2003b]. Control guidance sheet G202: laminar flow booth. In: COSHH essentials: easy steps to control chemicals. London: Health and Safety Executive.

58. HSE [2003c]. Control guidance sheet G206: sack filling. In: COSHH essentials: easy steps to control chemicals. London: Health and Safety Executive.

59. HSE [2003d]. Control guidance sheet G208: sack emptying. In: COSHH essentials: easy steps to control chemicals. London: Health and Safety Executive.

60. HSE [2004]. Nanoparticles: an occupational hygiene review. By Aitken RJ, Creely K S, Tran CL. London: Health and Safety Executive, Health & Safety Executive Publication No. RR 274.

61. Huang RF, Wu YD, Chen HD, Chen CC, Chen CW, Chang CP, Shih TS [2007a].

62. Development and evaluation of an air-curtain fume cabinet with

considerations of its aerodynamics. Ann Occup Hyg 51(2):189–206.

63. Huang S-H, Chen C-W, Chang C-P, Lai C-Y, Chen C-C [2007b]. Penetration of 4.5 nm to aerosol particles through fibrous filters. J Aerosol Sci 38(7):719–727.

64. IFA [2009]. Criteria for assessment of the effectiveness of protective measures [http://www. dguv.de/ifa/en/fac/nanopartikel/ beurteilungsmassstaebe/index.jsp]. Date accessed: October

65. 18, 2012.

66. ISO [2007]. Workplace atmospheres—ultrafine, nanoparticle and nano-structured aerosols. Inhalation exposure characterization and assessment. Geneva, Switzerland: International Organization for Standardization, Publication No. ISO/TR 27628:2007.

67. ISO [2008]. Nanotechnologies—Health and safety practices in occupational settings relevant to nanotechnologies. Geneva, Switzerland: International Organization for Standardization, Publication No. ISO/TR 12885:2008.

68. Johnson DR, Methner MM, Kennedy AJ, Steevens JA [2010]. Potential for occupational exposure to engineered carbon-based nanomaterials in environmental laboratory studies. Environ Health Perspect 118(1):49–54.

69. Kim CS, Bao L, Okuyama K, Shimada M, Niinuma H [2006]. Filtration efficiency of a fibrous filter for nanoparticles. J Nanopart Res 8:215–221.

70. Kim SC, Harrington MS, Pui DYH [2007]. Experimental study of nanoparticles penetrations through commercial filter media. J Nanopart Res 9:117–125.

71. Kletz T [2001]. An engineer's view of human error. New York: Taylor & Francis, p. 296.

72. Lee JH, Kwon M, Ji JH, Kang CS, Ahn KH, Han JH, Yu IJ [2011]. Exposure assessment of workplaces manufacturing nanosized TiO2 and silver. Inhal Toxicol 23(4):226–236.

73. Lee JH, Lee SB, Bae GN, Jeon KS, Yoon JU, Ji JH, Sung JH, Lee BG, Yang JS, Kim HY, Kang CS, Yu IJ [2010]. Exposure assessment of carbon nanotube manufacturing workplaces. Inhal Toxicol 22(5):369–381.

74. Lee KW, Liu BYH [1980]. On the minimum efficiency of the most penetrating particle size for fibrous filters. J Air & Waste Manage. Assoc. 30(4): 377-381.

75. Lindeløv JS, Wahlberg M [2009]. Spray drying for processing of nanomaterials. J Phys: Conference Series 170(1).

76. Macher JM, First MW [1984]. Effects of air flow rate and operator

activity on containment of bacterial aerosols in an class II safety cabinet. Appl Environ Microbiol 48:481–485.

77. Maidment SC [1998]. Occupational hygiene considerations in the development of a structured approach to select chemical control strategies. Ann Occup Hyg 42(6):391–400.

78. Mark D [2007]. Occupational exposure to nanoparticles and nanotubes. In: Hester RE, Harrison RM, eds. Nanotechnology: consequences for human health and the environment. London: RSC Publishing, pp. 50–80.

79. Marzal F, Gonzalez E, Minana A, Baeza A [2003a]. Methodologies for determining capture efficiencies in surface treatment tanks. Am Ind Hyg Assoc J 64(5):604–608.

80. Marzal F, Gonzalez E, Minana A, Baeza A [2003b]. Visualization of airflows in push-pull ventilation systems applied to surface treatment tanks. Am Ind Hyg Assoc J 64(4):455–460.

81. Maynard AD [2007]. Nanotechnology: the next big thing, or much ado about nothing? Ann Occup Hyg 51(1):12.

82. McKernan JL, Ellenbecker MJ [2007]. Ventilation equations for improved exothermic process control. Ann Occup Hyg 51(3):269–279.

83. Methner M [2008]. Engineering case reports: effectiveness of local exhaust ventilation (LEV) in controlling engineered nanomaterial emissions during reactor cleanout operations. J Occup Environ Hyg 5(6):D63–D69.

84. Methner M, Hodson L, Dames A, Geraci C [2010]. Nanoparticle emission assessment technique (NEAT) for the identification and measurement of potential inhalation exposure to engineered nanomaterials—part B: results from 12 field studies. J Occup Environ Hyg 7(3):163–176.

85. Methner MM, Birch ME, Evans DE, Ku BK, Crouch K, Hoover MD [2007]. Case

86. study: identification and characterization of potential sources of worker exposure to carbon nanofibers during polymer composite laboratory operations. J Occup Environ Hyg 4(12):D125–130.

87. Mukherjee SK, Singh MM, Jayaraman NI [1986]. Design guidelines for improved water spray systems. Min Eng 38(11):1054–1059.

88. Nanocyl [2009]. Responsible care and nanomaterials case study. Paper presented at the European Responsible Care Conference, Prague, October 21–23.

89. Naumann BD, Sargent EV, Starkman BS, Fraser WJ, Becker GT, Kirk GD [1996].

90. Performance-based exposure control limits for pharmaceutical active ingredients. Am Ind Hyg Assoc J 57(1):33–42.

91. NIOSH [1997]. Control of dust from powder dye handling operations. Cincinnati: U.S. Department of Health and Human Services, Centers for Disease Control and Prevention, National Institute for Occupational Safety and Health, DHHS (NIOSH) Publication

92. No. 97–107.

93. NIOSH [2004]. NIOSH respirator selection logic. Cincinnati, Ohio: U.S. Department of Health and Human Services, Centers for Disease Control and Prevention, National Institute for Occupational Safety and Health, DHHS (NIOSH) Publication No. (NIOSH) 2005 100.

94. NIOSH [2009a]. Approach to safe nanotechnology: managing the health and safety concerns associated with engineered nanomaterials. Cincinnati, OH: U.S. Department of Health and Human Services, Centers for Disease Control and Prevention, National Institute for Occupational Safety and Health, DHHS (NIOSH) Publication No. 2009–125.

95. NIOSH [2009b]. Qualitative risk characterization and management of occupational hazards: control banding (CB)—a literature review and critical analysis. Cincinnati, Ohio: U.S. Department of Health and Human Services, Centers for Disease Control and Prevention, National Institute for Occupational Safety and Health, DHHS (NIOSH) Publication No. 2009–152.

Chapter 3

DESIGN AND FABRICATION OF REMOTE WELDING EQUIPMENT IN A HOT-CELL

Soosung Kim,[1] Kihwan Kim,[1] Jungwon Lee,[1] and Jinhyun Koh[2]

[1]Research Reactor Fuel Development Division, Korea Atomic Energy Research Institute, Yuseong-gu, Daejeon 305-353, Republic of Korea

[2]Korea University of Technology and Education, Cheonan 330-708, Republic of Korea

ABSTRACT

The remote welding equipment for nuclear fuel bundle fabrication in a hot-cell was designed and developed. To achieve this, a preliminary investigation of hands-on fuel fabrication outside a hot-cell was conducted with a consideration of the constraints caused by the welding in a hot-cell. Some basic experiments were also carried out to improve the end-plate welding process for nuclear fuel bundle fabrication. The resistance welding equipment using end-plate welding was also improved. It was found that the remote resistance welding was more suitable for joining an end-plate to end caps in a hot-cell. This paper presents an outline of the developed welding equipment for nuclear fuel bundle fabrication and reviews a conceptual design of remote welding equipment using a master-slave manipulator. Furthermore, the mechanical considerations and a mock-up simulation test were described. Finally, its performance test results were presented for a mock-up of the remote resistance welding equipment for nuclear fuel bundle fabrication.

INTRODUCTION

Fuel cycle technology is being developed at KAERI and is meant to reuse spent PWR (Pressurized Water Reactor) fuel as raw material for the CANDU reactor [1]. This technology is being developed in the IMEF (Irradiated Material Examination Facility) at KAERI because of the nature of the high radioactivity of spent PWR fuel. In order to fabricate DUPIC (Direct Use of PWR fuel In CANDU reactors) nuclear fuels in a hot-cell environment, remote welding technology should be employed to weld between an end-plate and an end cap. To achieve this, a preliminary investigation of a hands-on fuel fabrication outside a hot-cell was necessary in consideration of the constraints caused by remote welding in a hot-cell [2]. The DFDF (DUPIC Fuel Development Facility) is a completely shielded cell made of heavy concrete. As the DFDF is active, direct human access to its inner cell is not possible. All the nuclear fuel fabrication processes and equipment operations, therefore, are conducted in a fully remote manner, using a master-slave manipulator.

In order to select a more suitable welding process in a hot-cell environment, various welding processes such as GTAW (Gas Tungsten Arc Welding), RW (Resistance Welding) [3], and LBW (Laser Beam Welding) methods, which are now available for end-plate welding for the commercial fuel bundle fabrication, should be processed as candidates. Even though the GTAW process is widely used for fabricating fuel bundles, it cannot be recommended for remote end-plate welding of a nuclear fuel bundle in a hot-cell facility due to the complexity of electrode alignment and difficulty in parts replacement in a remote manner. On the other hand, the RW process has some advantages because it is a qualified process and is extensively used in production. Hence, remote welding equipment was needed in order to join end-plates to end caps in a remote manner. The objective of this paper is to present the development of nuclear fuel bundle welding equipment for use in the highly radioactive zone of the DFDF at KAERI.

DESIGN OF REMOTE WELDING EQUIPMENT

All equipment for the remote fabrication of the nuclear fuel elements will be installed in a hot-cell as shown in Figure 1. As long as commercial equipment is available, it is purchased and modified for hot-cell use from the viewpoint of easy remote operation and maintenance. Among them, the remote welding equipment was designed to be modular and remotely operable by using a master-slave manipulator as shown in Figure 1(No. 21). The main parts of the welding equipment are located inside a hot cell, while the electronic parts are separated for installation outside a hot cell. The remote welding equipment is developed by adopting a head torch in order to achieve spot weld metal between an end-plate and end caps using a master-slave manipulator. As for end-plate welding operation and handling in a remote manner as shown in Figure 2, the design concept should take into account the remote manipulation, welding procedure, and capabilities and constrains of the remote handling devices that are available in a hot-cell. The design should also include considerations of an interface with a human operator, modular assembling parts for easy maintenance, electrical power transmission for control, and radiation effects of the materials to be used.

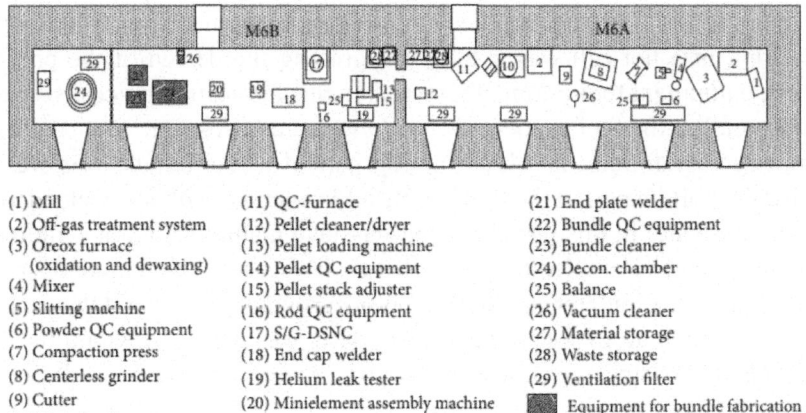

(1) Mill
(2) Off-gas treatment system
(3) Oreox furnace
      (oxidation and dewaxing)
(4) Mixer
(5) Slitting machine
(6) Powder QC equipment
(7) Compaction press
(8) Centerless grinder
(9) Cutter
(10) Sintering furnace

(11) QC-furnace
(12) Pellet cleaner/dryer
(13) Pellet loading machine
(14) Pellet QC equipment
(15) Pellet stack adjuster
(16) Rod QC equipment
(17) S/G-DSNC
(18) End cap welder
(19) Helium leak tester
(20) Minielement assembly machine

(21) End plate welder
(22) Bundle QC equipment
(23) Bundle cleaner
(24) Decon. chamber
(25) Balance
(26) Vacuum cleaner
(27) Material storage
(28) Waste storage
(29) Ventilation filter
▨ Equipment for bundle fabrication

Figure 1: Layout of the remote process equipment for nuclear fuel fabrication in DFDF.

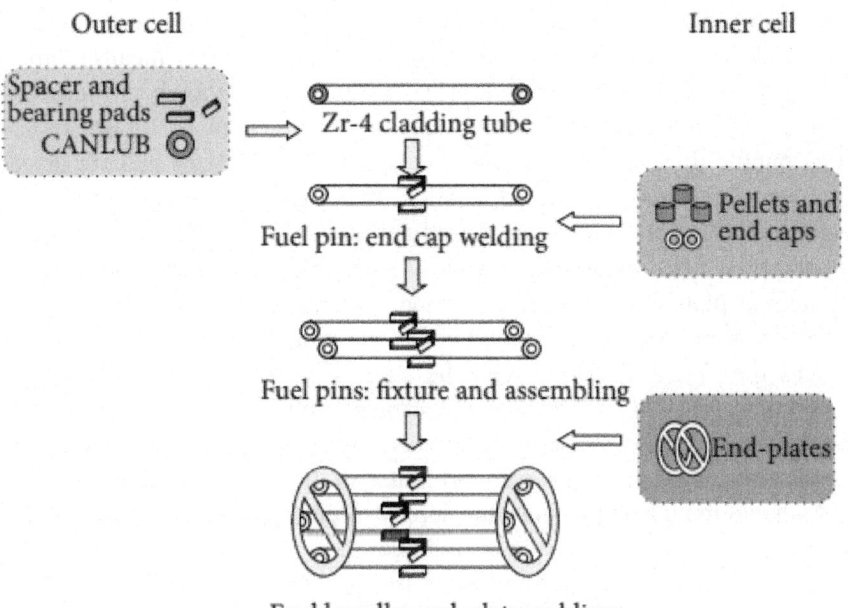

Figure 2: Process of end-plate welding operation and handling.

The remote welding system of the hot-cell environment consists of a resistance welding equipment, master-slave manipulator, and controller. The main head of the welding equipment will be used through a multi-pulse type method [4]. The modular remote welding equipment for a fuel bundle fabricating in a hot-cell was made by upgrading the design of the previous welding equipment for multipin fuel types. In this fabricating process sequence of fuel bundles, the fuel elements welded by the end caps were first positioned in an assembly fixture, in which the top part of a fuel bundle was welded. Finally, the bottom part of the fuel bundle after rotating 180° was welded to the bottom end-plate. In this process, a master-slave manipulator was required to be designed and assembled to be handled easily because the remote operation using a slave manipulator in a hot-cell was carried out. The modular welding equipment made up of four subassembly parts was designed with a modular concept and is compact in comparison with a previous welder for multipin fuel types in a remote manner.

The remote welding equipment consists of a main frame, a weld head using a single electrode, a branch electrode indexer, an endplate magazine loader, and a bottom assembler [5–7]. Figure 3 show the basic concept of the welding equipment, and Figure 4 illustrates the design construction of the remote welding equipment. The base frame itself consists of a single W-Cu electrode, a step-down transformer, an air cylinder, or other means of applying a change of the W-Cu electrode using a head pin as shown in Figure4(a). A branch electrode indexer provides accurate rotation of the upper and lower fuel bundle during end-plate welding operations. A rotary indexer driven by the servo motor is adjustable to allow the length of the overall shafting to vary as the indexing units are raised and lowered. The shafting for a remote operation is fitted together by means of a linear guide and linear bearing slides. A jigging plate using the Be-Cu branch electrodes, as shown in Figure 4(a), provide an accurate seat for the bundle end-plates and 37 elements. This part aligns the W-Cu electrode with the ends of 37 elements during a welding operation. End-plate loading mechanisms were used for the upper and lower units. An end-plate magazine loader as shown in Figure 4(a), dispenses and loads either the upper or lower end-plates to the bundle welding operation. A reloadable magazine provides the supply of end-plates to the units, which are dispensed one at a time by an air cylinder. A tuner unit of a bottom assembler was incorporated into the end-plate transfer gripper tooling to execute the rotation of the end-plate required during transfer, as shown in Figure 4(b). This unit is very robust, and thereby adheres to the permissible load restrictions of requiring no maintenance. Each of these subassembly parts of the remote welding equipment was designed in modules to facilitate maintenance by remote manipulation.

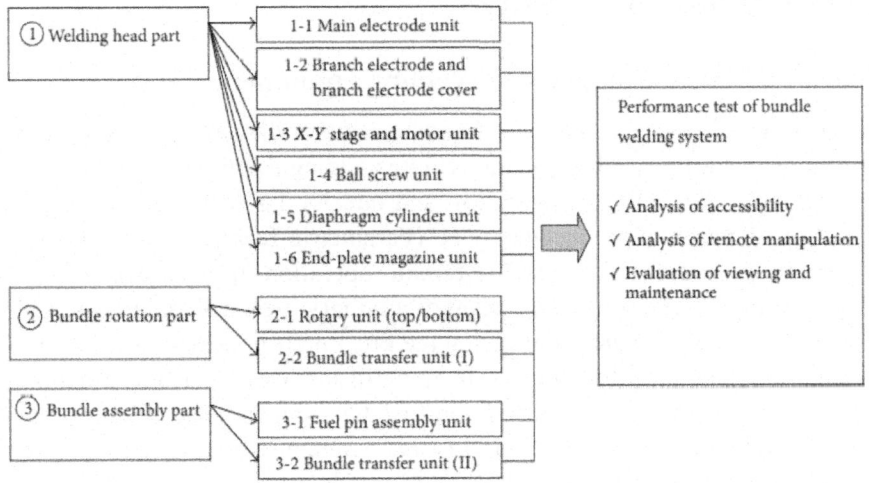

Figure 3: Basic concept of the remote welding equipment.

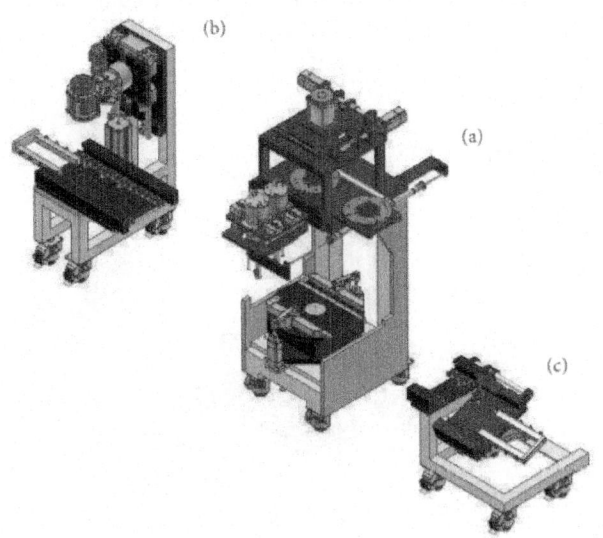

(a) Weld head part:
    Main electrode unit
    Main frame unit
    Branch electrode unit
    X-Y stage unit
    End-plate magazine unit

(b) Bundle rotation part:
    Bundle rotation unit
    Vertical transfer unit

(c) Bundle assembly part:
    Fuel elements assembly
    Angle control unit
    Bundle transfer unit

Figure 4: Design construction of remote welding equipment.

RESULT AND DISCUSSION

Mock-up Tests Using Remote Welding Equipment

After completing a basic drawing for the remote welding equipment, as shown in Figure 5, a design of the modular remote welding equipment was conducted by making a structural configuration and developing with the Pro-Engineer Wildfire 3.0 program produced by PTC (Parametric Technology Corporation). Based on the modular design, the remote operation in a hot-cell using the manipulator was checked with the aid of auxiliary exploded and re-assembled functions along with animations using the Pro-Engineer design method shown in Figure 6. The installing and exchanging of main parts such as a damaged weld head using a W-Cu electrode and Be-Cu branch electrodes for remote operation in a hot-cell, as shown in Figure 7, were also checked and analyzed. All the modular components of the assembling parts can also be remotely exchanged or maintained.

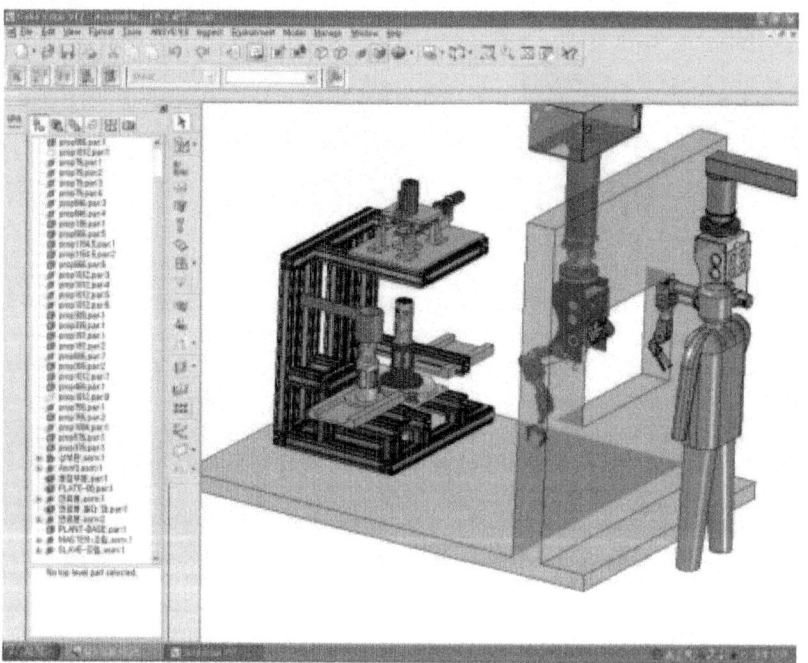

Figure 5: Schematic of remote welding system using by Pro-Engineer design method.

In order to prevent the weld head part from dropping from the master-slave manipulator during remote operation, the weld head part uses an assistant gripper. The assistant gripper is connected by using the crane of the roof door. As with real operation, the maximum weight that the master-slave manipulator can grasp is about 6 kg, at which point the operator can feel that the weight of the weld head part is about 1.7 kg. A mock-up simulation test was also carried out to check technological issues for remote operation and each element for the processing sequence. It was confirmed that the mock-up simulation test showed the process sequence and remote welding operation in a hot-cell environment using animation with the Pro-Engineer design method.

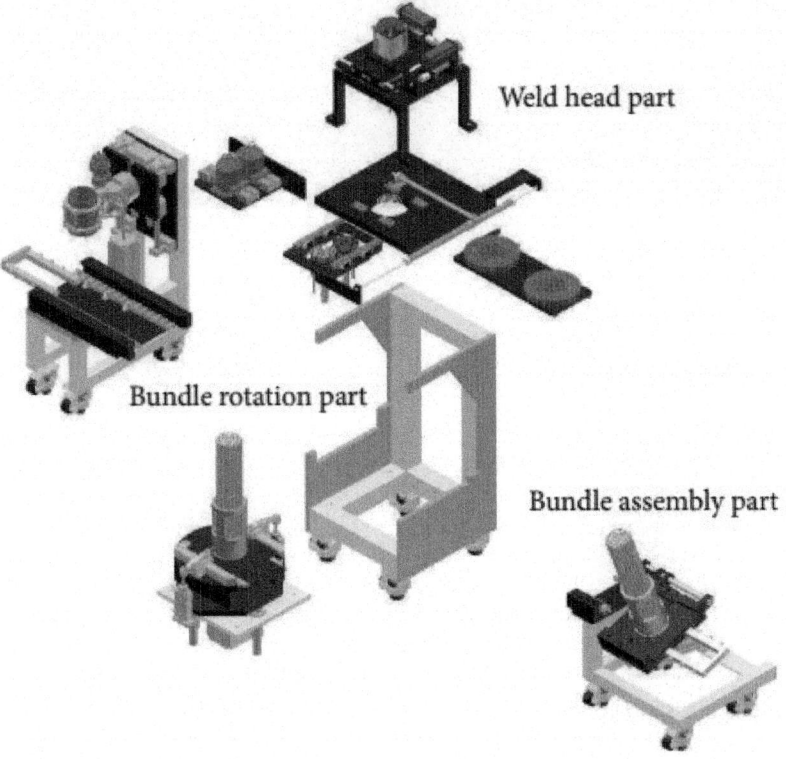

Figure 6: Auxiliary exploded and reassembled functions of top assembler.

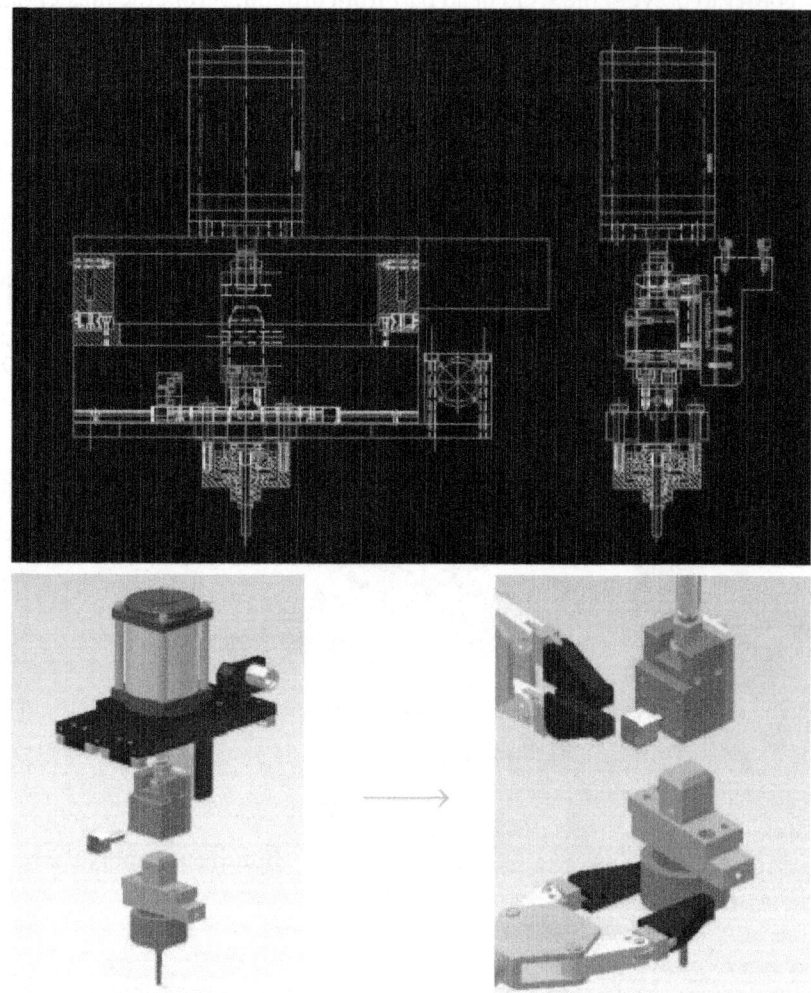

Figure 7: Illustration of designing, installing, and exchanging weld head.

In order to investigate weld performance of the fabricated welding equipment in a mock-up test room, a special fixture for a weld performance test was designed and fabricated as shown in Figure 8. These weld performance tests using a special fixture were performed according to the quality control procedure (Doc. No. HQP-33-02), and the weld samples were provided for four specimens (inner and outer rod specimens No. = 12, 16, 31, 37) [8]. The torque strengths of the weld performance test were created by the average values of four weld specimens as shown in Figure 9. In the real experiment, a special fixture was installed on the assembly floor of the remote welding equipment and the operator performed the welding operation by handling the master slave manipulator. This experiment was carried out by varying the working parameters of the weld current and the pressure of the main electrode, while the welding sequence controlled by the operator is constant [9]. Table 1 shows the welding conditions for the weld performance test and experimental results. The experimental results show that the weld current and pressure of main electrode influence the extent of the torque strength. Figure 9 shows the relationship between the weld current and torque strength using the Zircaloy-4 weld specimens. As for the effect of the torque strength during the welding operation, it was found that the torque values increased by increasing the weld current. Figure 10 shows the relationship between the pressure of the main electrode and torque strength of the Zircaloy-4 welded specimens at a current of 4200 A. To the extent of 3.5 BAR to 5 BAR of the pressure of the main electrode, the torque strengths of Zircaloy-4 welded specimens are found to be approximately 13-14 Nm, which is larger than that of the acceptable criteria (9 Nm) as followed by the quality control procedure. It is concluded from the experiment that for the welding conditions of the weld current and pressure of the main electrode, the torque strength reaches 99.9% in accomplishing perfect values of the resistance of Zircaloy-4 welded specimens. Figure 11 shows a photomicrograph of typical weld nuggets of the end-plate joint at magnification of X15. The lower portions of the micrograph show the weld zones of end cap joint which consist of inner rod and outer rod, welded to the end-plate joint for fuel element fabrication. The microstructure of a weld nugget consisted of a few re-crystallized and a little grown grains around an interface depending upon the heat cycles reached during the resistance welding. The weld metal of an end-plate region as shown in Figure 12 had a Widmanstätten structure with nonparallel α plates and was a basket-weave type [10]. The weld metal of an end cap region was also composed of a mixed structure of nonparallel Widmanstätten plates and quenched martensitic structures due to the very rapid heating and cooling cycle of the resistance welding [10, 11].

Table 1: Resistance welding conditions of the weld performance test

Sample no.	Weld current (A)	Pressure of main electrode (Bar)	Pressure of branch electrode (Bar)
# 01	4200	3.5	5.5
# 02	4200	4.0	5.5
# 03	4200	4.5	5.5
# 04	4200	5.0	4.5

Figure 8: Photograph of the welding fixture using Zircaloy-4 weld specimen.

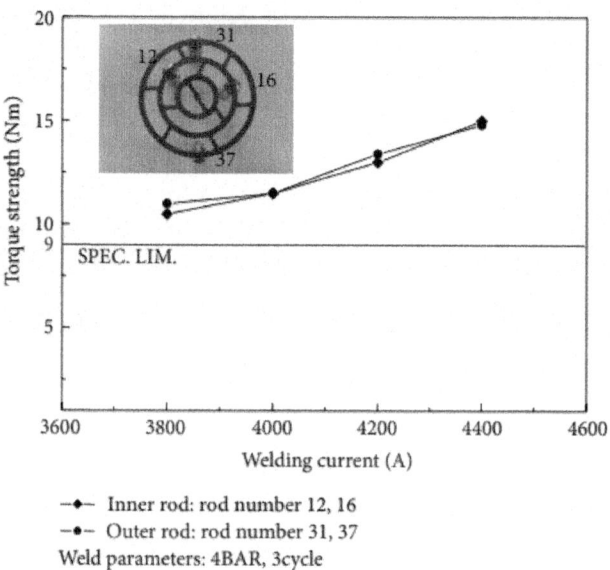

—♦— Inner rod: rod number 12, 16
—●— Outer rod: rod number 31, 37
Weld parameters: 4BAR, 3cycle

Figure 9: Torque strength of the welded specimens as a function of the weld current.

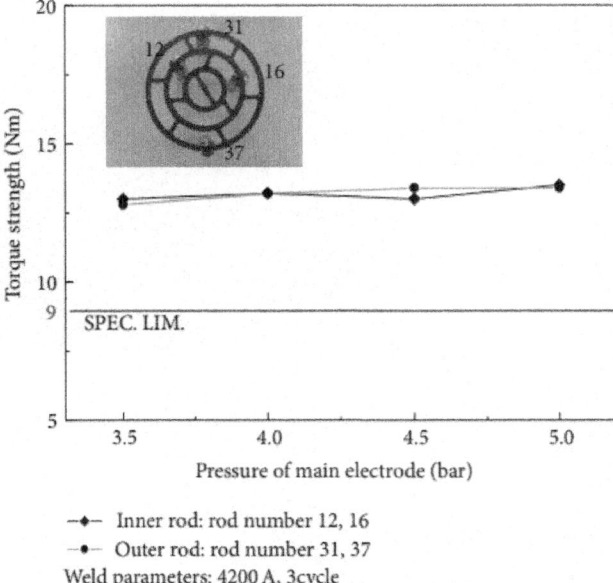

—◆— Inner rod: rod number 12, 16
—•— Outer rod: rod number 31, 37
Weld parameters: 4200 A, 3cycle

Figure 10: Torque strength of the welded specimens as a function of the pressure of main electrode.

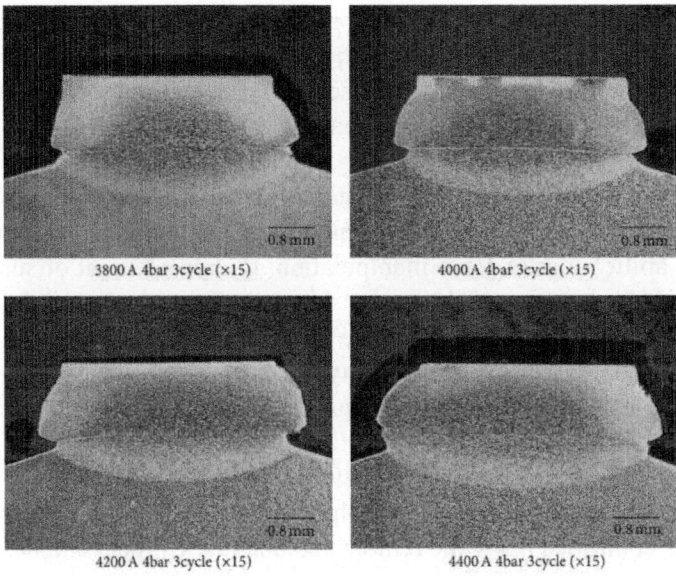

Figure 11: Macrocross sections of the nugget welds in end-plate to end cap joint (×15).

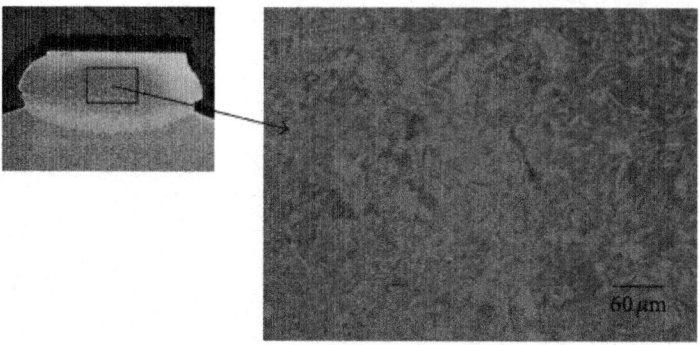

Figure 12: Microstructure of the central region of weld metal corresponding to part of the end-plate (×500).

Simulation Tests in Mock-up Facility

The remote welding equipment has been tested to verify its performance and capabilities in a mock-up of the simulation test room. The mock-up test environment of a remote welding equipment for the in-cell operation is shown in Figure 13. The human operator and manipulation are located outside of the mock-up of the simulation test room, and remote welding equipment is located inside it. Through the design and fabrication of remote welding equipment, Figure 13 shows combined welding equipment including the main frame part, the rotation part and the element assembly part in the mock-up simulation test room. Among them, the weld head of the main frame part is designed, and there is a reason why it was very interchangeable. Figure 14 illustrates remote operations using the master-slave manipulator. The mock-up performance test using remote welding equipment shows satisfactory results in access ability, master-slave manipulation, the replacement of subassembly parts, and operation repeatability, as shown in Table 2. Currently, the weld head and branch electrode parts are under performance tests in order to determine their reliability and stability before put into the DFDF. From the mock-up test, the remote welding equipment is improved and implemented to verify its performance and capabilities of all assembly parts. The mock-up tests including functional connections of the welding equipment for the bundle end-plate welding operation where a real operation though the hot-cell window is controlled using the remote and automatic process mode, are under development at the simulation test room.

Table 2: Results of a performance test for a remote welding equipment

Main parts	Subassembly parts	Access ability	Master-slave manipulation	Replacement of sub-parts	Repeatability (technical lessons*)
			Performance tests		
(1) Head part	1-1 main electrode unit	O	O	O	O
	1-2 branch electrode unit	O	O	O	O
	1-3 X-Y stage unit	△	△	△	△
	1-4 servo-motors	△	△	△	△
	1-5 ball screw unit	△	△	△	△
	1-6 diaphragm cylinder	△	△	△	△
	1-7 end-plate magazine box	△	△	△	△
(2) Rotation part	2-1 bundle rotary unit	O	O	O	O
	2-2 bundle transfer unit (I)	O	O	O	O
(3) Elements assembly part	3-1 fuel elements assembler	O	O	O	O
	3-2 bundle transfer unit (II)	O	O	O	O

* This is to be required in order to acquire its reliability and stability of remote welding equipment.
O: (Good), △: (Medium), ×: (Bad).

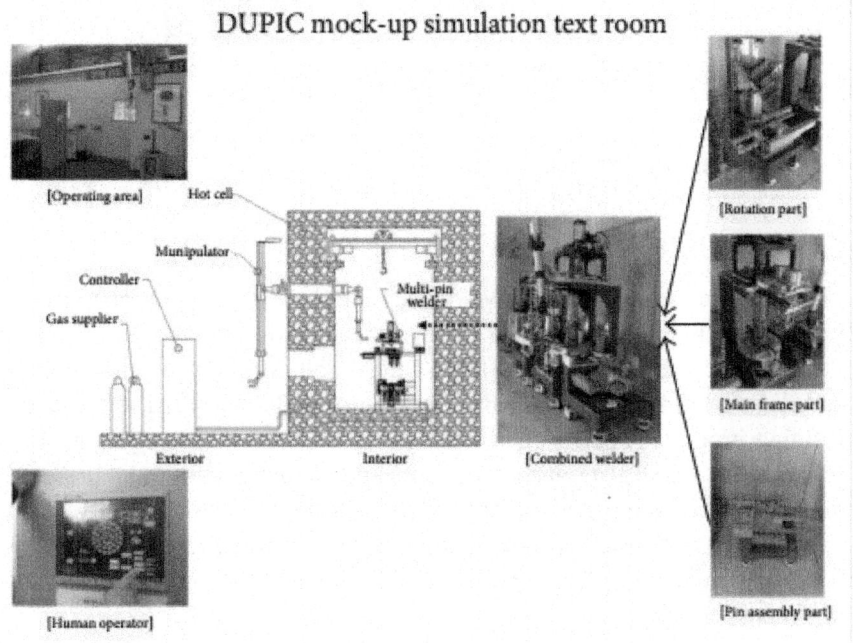

Figure 13: Remote welding equipment and its operation in mock-up facility.

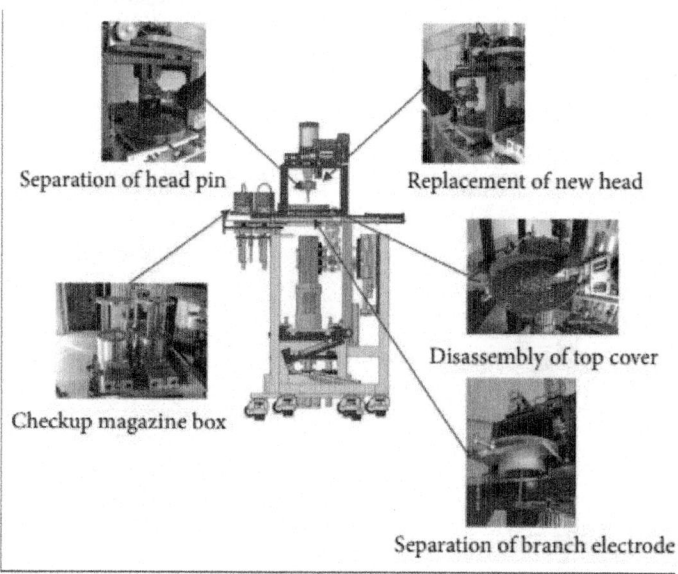

Figure 14: Illustration of manipulating various main parts.

CONCLUSION

This work was conducted to develop the remote welding equipment for nuclear fuel bundle manufacturing and to review the basic drawing by means of the Pro-Engineer design method. In the future, the optimum welding equipment and detailed drawings obtained in this study will be applied to the end-plate welding process. A performance test using remote welding equipment shows satisfactory results in access ability, master-slave manipulation, and the replacement of some parts. Furthermore, to establish the reliability of remote operations using resistance welding equipment, it was necessary to carry out a welding sample test using the end-plates for nuclear fuel bundle manufacturing in a mock-up facility.

ACKNOWLEDGMENTS

This research was supported by the Development of key technologies for Research Reactor Fuel Development sponsored by the Ministry of Education, Science and Technology (MEST) and by a grant from the Fundamental R&D program from Core Technology of Materials Funded by the Ministry of Knowledge Economy (MKE), Republic of Korea.

REFERENCES

1. M. S. Yang, et al., DUPIC Manufacturing and Process Technology, KAERI-1022/99, 2000.

2. J. D. Sullivan, The 10th KAIF/KNS Annual Conference, 1995.

3. L. M. Gourd, Principles of Welding Technology, Edward Arnold Publishers, 1980.

4. American Welding Society (AWS), Welding Handbook, vol. 3, 9th edition, 2007.

5. M. Malberg and N. Bay, "Methods of characterizing electrical systems of resistance welding machines," Welding Journal, vol. 77, no. 4, pp. 59–62, 1998. View at Google Scholar · View at Scopus

6. R. J. Bowers, C. D. Sorensen, and T. W. Eagar, "Electrode geometry in resistance spot welding,"Welding Journal, vol. 69, no. 2, p. 45S, 1990. View at Google Scholar

7. GE Canada Nuclear Products: Bundle Assembly Welder Manual, KNFC Equipment Data Book, 1995.

8. S.-S. Kim, G.-I. Park, J.-W. Lee, and J.-H. Koh, "Development of a remote welding machine for a dupic fuel bundle fabrication," in Proceedings of the 17th International Conference on Nuclear Engineering (ICONE '09), pp. 1–8, bel, July 2009. View at Scopus

9. S. S. Kim, J. W. Lee, G. I. Park, and J. H. Koh, "Development of zircaloy-4 endplate welding technology for a DUPIC fuel bundle assembly," Journal of Nuclear Science and Technology, vol. 46, no. 2, p. 103S, 2009. View at Publisher · View at Google Scholar

10. R. A. Bordoni and A. M. Olmedo, "Microstructure in the weld region in seam welded and resistance welded Zircaloy 4 tubing," Journal of Materials Science, vol. 16, no. 6, pp. 1527–1532, 1981. View at Publisher · View at Google Scholar · View at Scopus

11. R. A. Holt, W. Evans, and B. A. Cheadle, Role of Zirconium Alloy Metallurgy in the Fabrication of Candu Fuel, Atomic Energy of Canada Limited. AECL-5107, 1975. View at Scopus

Chapter 4

THE NEWEST GETTER TECHNOLOGIES: MATERIALS, PROCESSES, EQUIPMENT

Konstantin Chuntonov[1], Janez Setina[2], Gary Douglass[3]
[1]NanoShell Consulting, Nitzanim, Migdal Haemek, Israel
[2]Institute of Metals and Technology, Ljubljana, Slovenia
[3]Agile Chemistry, Inc., Elmhurst, USA

ABSTRACT

The efficiency of sorption purification of gases, as measured by an improvement in product quality and/or lowering of its cost, can be significantly increased via simple solutions: the substitution of current getter technology with reactive getters; and stimulation of the material in the sorption process using mechanochemical methods instead of heating or cooling. These ideas were embodied by the authors in new sorption apparatuses and devices such as mechanochemical sorption apparatuses for production of ultra pure gases, improved gas purifiers with reactive sorbent for production of pure and high purity gases and, finally, fluidized bed columns for mass production of pure and high purity gases.

INTRODUCTION

Big changes are brewing in the field of getter technologies: materials of more reactive nature (alloys of alkali- alkaline-earth and some rare-earth metals) claim the place of getters containing transition metals. These new gas sorbents do not need heating or cooling; they are more economical to use in production, more convenient in their operation and much more effective in sorption. These new materials outperform the current getter materials by orders of magnitude in sorption capacity [1] [2] .

Reactive getters can be used in place of all structural forms of sorption materials, such as films [3] -[14] , porous bodies and composites [15] -[22] , and powders [23] -[28] , i.e. everything that can be used in purification processes for flow gases, for capturing of moisture in multilayer OLED films, for capturing of residual gases in vacuum chambers, etc. It was found that, in the most common getter applications, the powder form of reactive alloy was preferable for economic and sorption reasons. Therefore, special efforts were made in recent years which were focused on solving the problem of reactive getter powders.

The result of the research and development which was performed was the creation of an entire family of new technologies united by a common sorption agent and represented currently by sorption columns of three types: mechanochemical sorption apparatuses [28] , gas purifiers using reactive sorbent and sorption columns of fluidized bed type [29] . This new equipment expands our practical possibilities, both in the studies of sorption processes and in their usage. The objective of this overview is to provide application specialists with the first knowledge about how the new technologies work and what can be expected from them.

MATERIALS—REACTIVE METAL POWDER

In the new technologies the gas impurity sorbents are alloys of reactive metals which sorb gases at room temperature. This is done by the growth of layers of chemical compounds on metallic surfaces. The choice of these types of new getter materials is easy to understand when they are compared to the sorption mechanism of the best known metal gas sorbents currently in use.

Figure 1 shows three types of sorption behavior of the metals which are used in getter technologies [2] [30] - [33] . Here Q is the amount of gas sorbed by a unit surface area by the moment of time t.

Curve 1 describes the sorption law at T_{room} for transition metals like Ti, V, Zr, Ni, Fe, etc., which serve the material basis of modern getter products. It is seen that sorption stops at t = t_p, when the surface is saturated with adatoms of gas. The ultimate value of $Q_1^* = k_1 h_p x_1$, where k_1 is the conversion factor and x_1 is concentration of gases in the passivated layer of thickness h_p answers this passivated state. The value of h_p is close to the thickness of a monoatomic layer.

Curve 2 describes the case of absorption: gases dissolve in the volume of the getter material. The value Q at absorption asymptotically approximates the value of Q_2^*, which is limited by the maximal solubility of gas in the metals of the given class. Ti, V, Nb and Zr behave in this way towards hydrogen at T_{room}, but for sorption of other gases these metals require heating. The ultimate value of specific sorption capacity at absorption can be found from the expression

$Q_2^* = k_1 h (Kp)^{1/2}$, where h is typical sorption size of the getter body (the thickness of the film, the radius of a spherical particle or a cylindrical needle, etc.), K is an equilibrium constant and p is partial pressure of gas molecules Y_2 above the getter.

Curves 3 and 4 refer to reactive metals which follow parabolic law $Q_3 = k_3 t^{1/2}$ (curve 3) or linear law $Q_4 = k_4 t$ (curve 4). This is the case of alkali, alkaline-earth and some rare-earth metals, which sorb target gases continuously to completion, until the entire material is consumed, as the result of the reaction Me + Y = MeY, where Me is metal, Y is sorbed gas, and MeY is the solid product of the reaction. Nitrogen sorption by barium films [5] can serve as a convincing illustration of a parabolic law; and oxygen sorption by lithium films [10] as an illustration of a linear law.

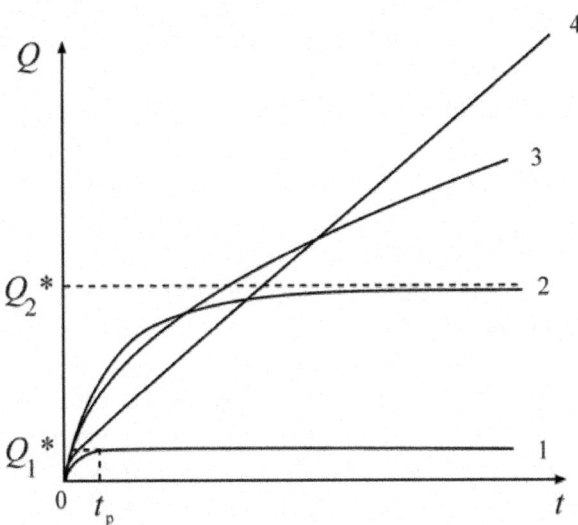

Figure 1: Sorption curves Q(t) and getter classes: 1: adsorption, 2: absorption, 3 and 4: chemical reactions with the frontal movement according to the parabolic or linear law correspondingly.

For the ultimate values of sorption capacities of reactive metals we can write: $Q_3^* = Q_4^* = k_1 h x_3$, where x_3 is the concentration of gas in MeY. Taking into consideration that $h_p \ll h$ and that the ultimate solubility of gas in metal at T_{room} is much lower than the concentration of this gas in the compound MeY, we come as the result to the correlation of the ultimate values

$$Q_4^* = Q_3^* > Q_2^* \gg Q_1^* ,$$

from which it is seen that at room temperature the sorption capacities of reactive metals (Q_4^* and Q_3^*) are by orders of magnitude higher than those of transition metals (Q_1^*) and their capacities are also higher than in the case of heating of the latter (Q_2^*).

The process of gas purification in flow apparatuses depends not only on the sorption capacity of the getter material but also on the sorption rate, which limits the allowable flow rate. For any sorption column the sorption rate can be written down as $J = (dQ/dt)S$, where dQ/dt is the specific sorption rate of the getter material and S is the surface area of this material available for target gases. The larger the sorbing surface area S the higher the sorption rate J; and this fact made us look for powder solutions, as it is powders which yield the maximal values of specific surface area. This explains why in our technologies the preference is given to reactive metals and why we prefer them in powder form.

The comparative sorption analogue of reactive powders from the side of the traditional getter materials are loose powders of transition metals or sintered powder materials on the basis of these metals in the form of pellets or high porous thick films [34] [35] . Figure 2 and Figure 3 show the sorption characteristics of three reactive alloys, intermetallic compounds of Ba_8Ga_7 and LiGa as well as eutectic alloy of Ba-20 at% In, these are compared to the established products of SAES Getters Inc. Measurements of the sorption properties were performed by dynamic flow method [36] [37] , where an apparatus with a disintegrator [28] , which allows mechanical milling of a reactive ingot in UHV conditions, served as a test chamber. Figure 2 shows the sorption curves of nitrogen, which is one of the most difficult target gases to remove. As it can be seen at room temperature the sorption capacity of high porous getter film St. 122 [38] is approximately 20 times lower than that of the powder LiGa and approximately 500 times lower than that of the powder Ba—20 at% In. The mass of the reactive powder in each of the cases was equal to 0.12 g and the particle size was in the range of from ~50 to 250 microns. Double-sided getter St. 122 had dimensions 2 cm × 3 cm with getter mass 0.24 g. It is quite clear that if in reactive alloys gallium or indium (which do not participate in nitrogen sorption) are replaced with a reactive metal, the sorption capacity of the getter powder will grow accordingly.

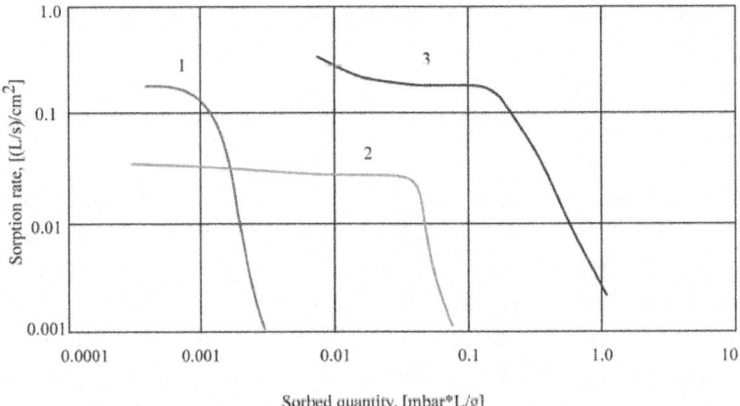

Figure 2: Sorption of nitrogen at room temperature: 1: St. 122 activated at 450°C in vacuum 5 × 10⁻⁷ mbar 1 hour; 2: powder of LiGa tested immediately after milling. 3: powder Ba-20 at% In tested 24 hours after milling.

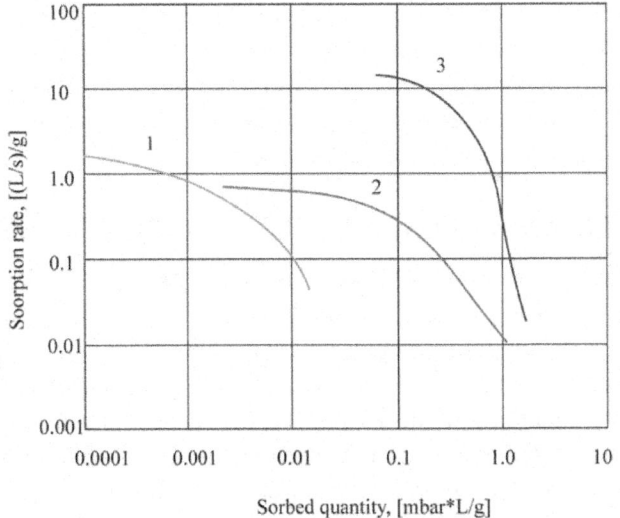

Figure 3: Sorption of oxygen and carbon oxides at room temperature: 1: St. 707 activated at 500°C for 10 minutes, test gas CO [39] ; 2: powder Ba_8Ga_7 tested 340 hours after milling, test gas O_2; 3: Ba-20 at% In tested 24 hours after milling, test gas CO_2.

Sorption characteristics of powders St. 707 as well as reactive powders of Ba-20 at% In and Ba_8Ga_7 are given in Figure 3. Oxygen and carbon oxides, which are very similar in their sorption behavior, were used as test gases. The mass of the powder of St. 707 was equal to 0.2 g and the particle size was 40 - 128 micron [39] ; the mass of each of the reactive powders was 0.25 g and the particle size was the same as in the case of the tests with nitrogen. According to the experimental data (Figure 3) the specific sorption capacity of the reactive materials towards oxygen and oxygen containing gases is on average approximately two orders of magnitude higher than that of the current getters.

The results obtained look impressive as the 100-fold advantage in sorption capacity immediately opens ways to industrial applications for reactive powders. With that said, the evaluation of the obtained values of sorption capacity of reactive metals shows that these values are far from the theoretical limit and that decreasing the average particle size to ~10 micron will allow an increase in the sorption capacity by 15 - 20 times more. Naturally, this fact identifies the need to improve the methods of the mechanical milling of reactive alloys. The solution can be reached by enhanced design or mechanical techniques and by influencing the structure of the ingot during the growth process.

We will now move from the materials overview to a comparative review of the sorption columns and the processes, which take place within them. Before we begin, let us remind the reader that, in contrast to the previous methods of gettering gases, the new technology does not require heating or cooling of the gas sorbent; in addition, it is by many times more effective in sorption while being significantly more economical to achieve.

IMPROVED GAS PURIFIERS WITH REACTIVE SORBENT

New gas purifiers of in-line type are intended for the production of pure and high pure gases within systems of small or medium capacity [29] . They work at T_{room}, have a typical look and are similar to currently used equipment in their operation; however, they are more convenient in operation, are completely safe and, most importantly, provide a 100-fold increase of sorption capacity within the same outside dimensions. The experimental results which were discussed in the previous chapter are primarily applicable to this type of gas purification equipment.

The design of the new gas purifier can be seen in Figure 4 [29] : its housing is formed by two blunted cones, which are put together along their common base forming divergent/convergent gas flow without "dead zones".

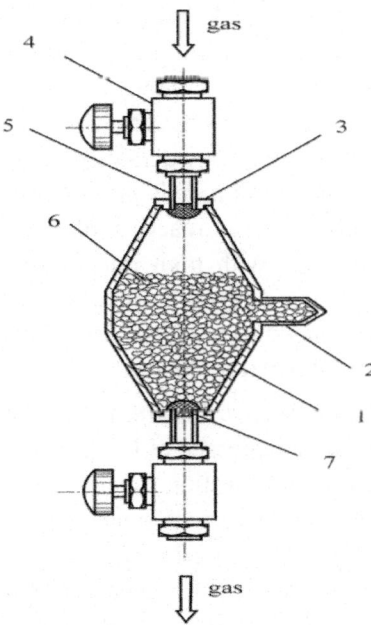

Figure 4: Gas purifier with reactive powder. 1: a steel housing with the remains of filling tube 2; 3: end ports; 4: a gas valve; 5: gas tubes; 6: sorbent particles; 7: a filter.

Free pouring of powder particles minimizes the pressure drop and two gas valves on the ends of the device make it autonomous. This allows its installation or a change of its working position without loss of the gas sorbent and without leakage of the process gas to the atmosphere. In a new purifier, the presence of the space not occupied by the powder inside the housing is necessary because the volume of the particles will increase and occupy the remaining space during the gas sorption process.

The material bases for sorption powders are binary, ternary or multicomponent alloys, the concentration of reactive metals in which the range is from 50% to 100%. It is under these conditions that gas sorption takes place according to the mechanism of chemical reactions and runs to completion. Among these alloys the preference is given to alloys of Ba-Al, Ba-Mg, as well as to alloys of Ca-Li, Ca-Mg or Ca-Li-Mg. It is not difficult to change the ratios of components in these alloys in order to create the required composition of getter material for purification of noble gases, nitrogen, hydrogen, etc.

The advantages of the improved gas purifiers are not solely limited to its sorption efficiency; they are attractive from the economical view point as well, due to the cheapness and availability of reactive metals. Market prices for reactive metals are by several times lower than for Ti, V and Zr. In the case of reactive powders the production costs to produce the end getter products from the initial metals are less expensive, by many times over. To illustrate, for the production of sintered powder porous materials, five technological operations are needed; in the case of reactive powders only two operations are needed. The creation of alloys of transition metals require high vacuum furnace with heating to 2000°C and higher, while reactive alloys are grown with the help of simple ampoule techniques with high rate in common tube furnaces at temperatures range from 600°C and lower. As a result, the cost for production of reactive powders is estimated to be approximately 10 times lower than for the production of the current getter materials.

If we define the total positive effect from the industry changeover to the new gas purifiers with a value expressed as $k = k_1 \times k_2$, where k_1 is the number showing by how many times the sorption capacity of reactive metals is higher than the sorption capacity of the current sintered powder porous getters and the number k_2 is the number showing by how many times production costs of sintered powder porous getters are higher than that of the production of reactive powders, we come to $k = 100 \times 10 = 1000$. This is a revolutionary result, especially taking into consideration that a decrease of the size of reactive particles to ~10 micron is able to lead to the increase of the total effect by an additional order of magnitude.

MECHANOCHEMICAL SORPTION APPARATUSES

These apparatuses were initially intended for the production of ultra pure gases [28] . They are new in principle equipment, which have no analogs in gas purification practice. Schematically such an apparatus is shown in Figure 5: The initial ingot loaded into disintegrator 2 and/or 2' and is milled directly in the atmosphere of purified gas X + Y, where X is the desired process gas, and Y is impurity. The gas, which enters the reactor through filter 4, goes then through a thin layer of as-made powder 12 and exits through filter 10.

The advantages of the mechanochemical method are described in detail in [27] [28] , so here we will mention only the primary advantages, which are the controlled character of the purification process and the clear absence of the factors limiting the purity degree of the end product.

Controllability of the sorption process: For the first time in sorption purification technologies a feedback system 6 (Figure 5) allowed making this process controllable and this is done in a simple way: by regulating the feed rate of getter powder into the reaction zone. By changing the amount of the powder mass produced during a unit of time it is possible to achieve constant purity of the gas, which exits column at the desired rate of gas flow.

Control over the purification process is a new useful function, which not only facilitates the standardization of related production processes but is also capable of bringing reorganization in the gas supply systems. The users of ultra pure gases can now look forward to significant reduction or elimination of certain logistical issues associated with these gases by building mechanochemical apparatuses directly into their gas network.

Extremely high purity of the output process gas: The estimations based on the data from gas permeability of stainless steels and in the data from partial pressure of the products MeY near ambient temperature show that mechanochemical purification of substances by reactive metals is able to decrease the content of active and low active impurities in them to $\sim 10^{-15}\%$. At present neither the gas industry nor the researchers have been able to set these kinds of goals for themselves as they have been absolutely unobtainable.

What is it that radically shifts the level of the achieved purity of the gas product in the mechanochemical method? Let us point out two factors. The first one is that in mechanochemical purification, which is carried out at ambient temperature, the sources of contamination which appear in getter technologies using the heating of purification materials are absent. Raising the temperature, as it is known, stimulates reactions with the formation of a volatile by-product and increases diffusion mobility of gas species through the column wall.

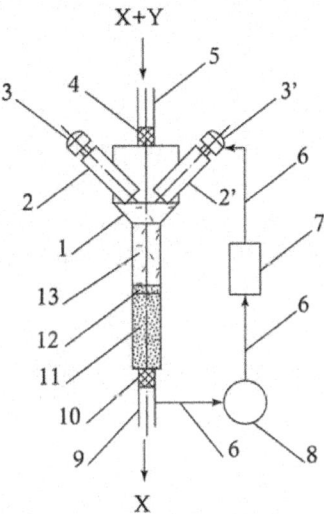

Figure 5: Mechanochemical sorption apparatus: 1: the impermeable to gas casing of the sorption column; 2 and 2': disintegrators; 3 and 3': outside actuators; 4: a filter; 5:

an inlet; 6: a feedback line; 7: a controller; 8: a gas analyzer; 9: an outlet; 10: a filter; 11: used powder; 12: a layer of fresh powder; 13: falling particles.

The second one is the super reactivity of as-milled solids. Powders produced by mechanical disintegration of solids in a cold state are oversaturated with defects. Activation barriers at this point abruptly decrease while the reactions and diffusion processes speed up. It should be also taken into consideration that reactive metals themselves already belong to the class of the most active chemical substances. That is, as-made powders of reactive metals and alloys are the best of what nature can offer us as purification material, of which all can be produced in the most economical manner. In fact, most of the above mentioned reactive metals are widely spread in the Earth crust, are easily available and are characterized by relatively low melting points.

Mechanochemical sorption apparatuses can be used not only for purification of flow gases [28] but also as vacuum getter pumps with controlled pumping speed, small weight and small energy consumption due to high fragility and low mechanical strength of the alloys [26] . This combination of characteristics makes the new getter pump attractable for employment in portable devices or in the measuring systems, where vibration is undesirable, e.g. in electronic microscopy.

One more application field for mechanochemical apparatuses is mechanochemistry itself, where our apparatuses are ready to become one of the instruments for research of mechanochemical reactions of gas/solid or liquid/solid type [27] [28] . Building up experimental dependences of z-t type, where z is the property under measurement and t is the time interval between the milling of the ingot and the beginning of the measurement, will allow the separation of contributions from different driving forces of the process under study. This will improve our understanding of the nature of such reactions. A reactor with a transparent wall will give a chance to visualize a lot of changes in the state of the system under study and use this information either for research purposes or for demonstration purposes in teaching the basics of mechanochemistry.

The general view of the disintegrator used for the production of reactive powders, the sorption characteristics of which are given in Figure 2 and Figure 3, is shown in Figure 6.

Figure 6: Vacuum milling chamber.

SORPTION COLUMNS OF THE FLUIDIZED BED TYPE

The third type of gas purification equipment working on metal reactive powders is schematically shown in Figure 7 [29] . The sorption column 1 reaches maximally high efficiency due to fluidized bed technology: here the inlet of the gas to be purified is located not from above but from below upwards through valve 3 and filter 2.

The regime of fluidized bed is set when the rate of the gas fed from below begins to exceed the incipient fluidization point [40] - [42] . From this moment the rising flows of gas X + Y set the particles of Me, which were earlier resting on the distributor plate 4, into chaotic movement. Colliding with each other, these particles release from the growing layer of products MeY on their surface, and this maintains the reaction Me + Y = MeY on a higher kinetic level than in gas purifiers of in-line type, where the rate is limited by the diffusion of the reagents through the layer of product MeY.

Small particles of MeY appearing as the result of chemical reactions and mechanical collisions in the fluidization zone are taken upwards by the gas flow, where the separation of the mixture MeY + X takes place: molecules of gas X passing through the exhaust filter 9, the outside filter 8, and the valve 7, and then leave the sorption column. Solid particles MeY then go through the discharge port 6 into the receiver of solid waste 5.

Sorption columns of the given type have the highest productive capacity compared to the previously discussed equipment and for this reason they are more suitable than the others for the mass production of pure and high purity gases. In contrast to the mechanochemical sorption apparatuses, where the reaction takes place in a thin layer of fresh powder, here the reaction involves the entire thickness of the material filled into the column. If the sorption columns are compared with the improved gas purifiers, they clearly show superiority in the process kinetics and are more convenient in industrial conditions because they do not need to be periodically replaced by replenished purifier components. Sorption columns (Figure 7), being the equipment of a stationary type, are replenished when necessary with fresh sorbent in their working position through the filling port 10 and the waste is separated from the system via the replaceable receiver 5.

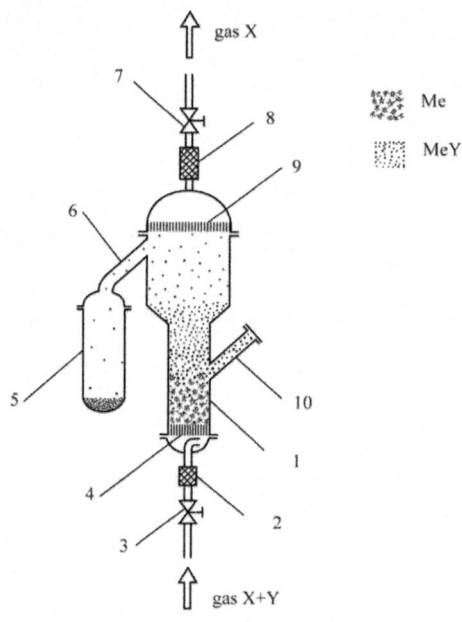

Figure 7: Sorption column of fluidized type: 1: a sorption column; 2: a filter; 3: a valve; 4: a distributer; 5: a receiver; 6: discharge port; 7: a valve; 8: an outside filter; 9: an exhaust filter; 10: a filling port.

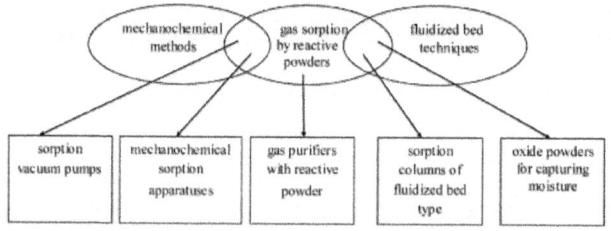

Figure 8: The structure of powder direction in getter technologies.

In addition to gas purification technologies, a sorption column of fluidized bed type can be used as a factory for the production of nano- or micro-particles of oxides of reactive metals, e.g. BaO or CaO, which are the best dryer materials. For the production of these kind of powders it is enough to blow argon with a small concentration of oxygen from below through the layer of the initial metallic particles and then small oxide particles, which are formed in the fluidized zone, are lifted by gas flow upward and are accumulated in receiver 5, which in this case becomes a collector of the end product.

CONCLUSIONS

Sorption apparatuses and devices with reactive metal powder are the new technological interdisciplinary directions in getters and gas purification, which appeared from the intersection of three fields: sorption of gases by reactive metals, mechanochemistry, and fluidized bed technique (Figure 8). In its content this direction does not go beyond the framework of the known getter applications; however, it radically improves the circumstances involved in their use. This takes place due to the improved sorption capacities and the improved economies associated with reactive metals, technological innovations and construction techniques, which allowed the creation of new gas purification equipment. The new reality in the getter field can now be fully understood taking into account the following developments:

- improved gas purifiers using reactive powder, the cost of which is by 10 times lower and the sorption capacity of which is by 100 times higher than those of the prototype using getters of a different class;

- mechanochemical sorption apparatuses, which for the first time performed operating control of the sorption process and purification of flow gas by mechanical activation of a reagent in situ and in statu nascendi;

- sorption columns of fluidized type for mass production of pure and high purity gases or for the production of powder drying materials of the composition BaO or CaO.

At present it is difficult to predict all the consequences of industrialization of these developments. Certainly a multiple increase in the life span of gas purification systems and decrease in their size will take place; gases of higher purity grade than the purity grade achieved for today will become available with a much lower cost stack than for the current product. With that said, however, the authors realize that despite the simplicity and obviousness of the suggested solutions it will take time until they are accepted by specialists in these fields, as this is connected with break of paradigms in the field, where nothing has essentially changed for a long time.

REFERENCES

1. Ferrario, B. (1996) Chemical Pumping in Vacuum Technology. Vacuum, 47, 363-370. http://dx.doi.org/10.1016/0042-207X(95)00252-9

2. Chuntonov, K. and Yatsenko, S. (2013) Getter Films for Small Vacuum Chambers. Recent Patent on Materials Science, Bentham Science Publishers, 6, 29-39. http://dx.doi.org/10.2174/1874464811306010029

3. Fransen, J.J.B. and Perdijk, H.J.R. (1960) The Absorption of Gases by Barium Getter Films Applied as a Tool. Vacuum, 10, 199-203. http://dx.doi.org/10.1016/0042-207X(60)90136-6

4. Ricca, F. and della Porta, P. (1960) Carbon Monoxide Sorption by Barium Films. Vacuum, 10, 215-222. http://dx.doi.org/10.1016/0042-207X(60)90140-8

5. della Porta, P. and Argano, E. (1960) Nitrogen Sorption by Barium Films. Vacuum, 10, 223-226. http://dx.doi.org/10.1016/0042-207X(60)90141-X

6. Turnbull, J.C. (1977) Barium, Strontium, and Calcium as Getter in Electron Tubes. Journal of Vacuum Science & Technology A, 14, 636-639. http://dx.doi.org/10.1116/1.569166

7. Verhoeven, J. and van Doveren, H. (1982) Interaction of Residual Gases with Barium Getter Film as Measured by AES and XPS. Journal of Vacuum Science & Technology, 20, 64-74. http://dx.doi.org/10.1116/1.571310

8. Sparks, D.R. (2005) Method of Forming a Reactive Material and Article Formed thereby. US Patent 6923625.

9. Silvernail, J.A. (2006) Protected Organic Electronic Device Structures Incorporating Pressure Sensitive Adhesive and Desiccant. US Patent No. 6998648.

10. Chuntonov, K. and Setina, J. (2008) New Lithium Gas Sorbents: I. The Evaporable Variant. Journal of Alloys and Compounds, 455, 489-496. http://dx.doi.org/10.1016/j.jallcom.2007.01.158

11. 11. Chuntonov, K., Setina, J., Ivanov, A. and Permikin, D. (2008) New Lithium Gas Sorbent: III. Experimental Data on Evaporation. Journal of Alloys and Compounds, 460, 357-362. http://dx.doi.org/10.1016/j.jallcom.2007.06.055

12. Chuntonov, K. (2011) Lithium or Barium Based Film Getters. US Patent Application 20110217491.

13. Horie, H., Fukuda, Y., Kato H., Nakashima, N. and Makino, Y. (2011) Getter Material and Evaporable Getter Device Using the Same, and Electron Tube. US Patent No. 7927167.

14. Giedraitis, A., Tamulevicius, S., Gudaitis, R. and Andrulevicius, M. (2010) Kinetics of Growth and Sorption Properties of Evaporable Barium Getter Films. Materials Science (Medziagotyra), 16, 12-23.

15. Alvarez Jr., D. (2001) Method and Apparatus for Purification of Hydride Gas Streams. US Patent No. 6241955.

16. Londer, H., Myneni, G.R., Adderley, P., Bartlok, G., Knapp, W., Schleussner, D. and Ogris, E. (2007) New High Capacity Getter for Vacuum Insulated Mobile LH2 Tank Systems. Vacuum, 82, 431-434. http://dx.doi.org/10.1016/j.vacuum.2007.07.063

17. Sparks, D.R., Najafi, N. and Newman, B.E. (2010) Getter Device. US Patent No. 7789949.

18. Giannantonio, R., Vescovi, C., Cattaneo, L. and Longoni, G. (2011) Getter Systems Comprising an Active Phase Inserted in a Porous Material Distributed in a Low Permeability Means. US Patent No. 7977277.

19. Giannantonio, R., Longoni, G., Vescovi, C. and Cattaneo, L. (2014) Getter System Comprising One or More Deposits of Getter Material and a Layer of Material for the Transport of Water. US Patent No. 8911862.

20. Chuntonov, K. (2013) Safe Gas Sorbents with High Sorption Capacity on the Basis of Lithium Alloys. US Patent No. 8529673.

21. Chuntonov, K. (2014) Barium Containing Granules for Sorption Applications. US Patent No. 8623302.

22. Chuntonov, K. (2014) Apparatus and Method for Droplet Casting of Reactive Alloys and Applications. US Patent Application 20140290897.

23. Boffito, C. and Schiabel, A. (1994) Process for the Sorption of Residual Gas by Means of a Non-Evaporated Barium Getter Alloy. US Patent No. 5312606.

24. Schiabel, A. and Boffito, C. (1994) Process for the Sorption of Residual Gas by Means of a Non-Evaporated Barium Getter Alloy. US Patent No. 5312607.

25. Manini, P. and Belloni, F. (1997) Device for Maintaining a Vacuum in a Thermally Insulating Jacket and Method of Making Such Device. US Patent No. 5600957.

26. Chuntonov, K. (2013) Sorption Pump with Mechanical Activation of Getter Material and Process for Capturing of Active Gases. US Patent Application 2013078113.

27. Chuntonov, K. and Lee, M.K. (2014) Mechanochemical Sorption Apparatuses. Advanced Materials Research, 875-877, 1106-1110. http://dx.doi.org/10.4028/www.scientific.net/AMR.875-877.1106

28. Chuntonov, K. (2015) Sorption Apparatus for the Production of Pure Gases. US Patent No. 9095805.

29. Chuntonov, K. and Setina, J. (2015) Activationless Gas Purifiers with High Sorption Capacity. US Patent Application (Patent Pending).

30. Kubaschewski, O. and Hopkins, B.E. (1962) Oxidation of Metals and Alloys. Butterworths, London.

31. Hauffe, K. (1966) Reaktionen in und an festen Stoffen. 2. Auflagen, Springer-Verlag, Berlin. http://dx.doi.org/10.1007/978-3-642-88042-1

32. Meyer, K. (1968) Physikalisch-chemische Kristallographie. VEB Deutscher Verlag, Leipzig.

33. Fromm, E. and Gebhardt, E. (1976) Gase und Kohlenstoff in Metallen. Springer-Verlag, Berlin. http://dx.doi.org/10.1007/978-3-642-80943-9

34. della Porta, P. (1992) "Gettering": An Integral Part of Vacuum Technology. Proceedings of the 39th National Symposium of American Vacuum Society, Chicago, 9-13 November 1992, Technical Paper TP 202.

35. Giorgi, E. and Ferrario, B. (1989) High-Porosity Thick-Film Getters. IEEE Transactions on Electron Devices, 36, 2744-2747. http://dx.doi.org/10.1109/16.43783

36. ASTM F 798-97 (2002) Standard Practice for Determining Gettering Rate, Sorption Capacity, and Gas Content of Non-Evaporable Getters in the Molecular Flow Region.

37. Erjavec, B. and Setina, J. (2011) Investigations of a Method for Determining Pumping Speed and Sorption Capacity of Nonevaporable Getters Based on in Situ Calibrated Throughput. Journal of Vacuum Science & Technology A, 29, Article ID: 051602. http://dx.doi.org/10.1116/1.3626535

38. Brochure SAES Getters (2004) Solution for Flat Panel Displays.

39. Toia, L. and Boffito, C. (2003) Non-Evaporable Getter Alloys. US Patent No. 6521014.

40. Gupta, C.K. and Sathiyamoorthy, D. (1999) Fluid Bed Technology in Materials Processing. CRC Press LLC, Boca Raton.

41. Zhu, J., Leckner, B., Cheng, Y. and Grace, J.R. (2005) Fluidized Beds. In: Crowe, C.T., Ed., Multiphase Flow Handbook, CRC Press, Boca Raton, 5-1-5-93http://dx.doi.org/10.1201/9781420040470.ch5

42. Epstein, N. and Grace, J.R. (2010) Spouted and Spout-Fluid Beds: Fundamentals and Applications. Cambridge University Press, Cambridge. http://dx.doi.org/10.1017/CBO9780511777936

Chapter 5

NEW TECHNOLOGY FOR GRIDS AND SCALES MANUFACTURING IN OPTICAL DEVICES

Vladimir Stepanovich Kondratenko, Vladimir Evgenievich Borisovsky, Alexandr Sergeevich Naumov, Nikolay Eduardovich Petruljanis

[1]International Graduate School of Business, University of South Australia, Adelaide, Australia

[2]Faculty of Management, Multimedia University, Cyberjaya, Malaysia

ABSTRACT

Using the laser controlled thermocracking method, research results for the new technology of optical grids and scales manufacturing are given in this paper. The opportunity of grids and scales manufacturing is shown for a wide range of the sizes, scale's pitches and its width: From 10 nanometers up to 10 microns with a backlight in various optical ranges.

INTRODUCTION

In the manufacture of many optical devices, one of the most important process steps is grids and scales making on this devices. Grid and scale are used to make the sampling, measurement and also for pointing the device at the subject.

Most of the grids and scales are applied for the parallel sided plate's surface, which are predominantly placed in the focal plane of the oculars of the optical system [1].

Depending on the tolerances for linear and angular dimensions, grids are subdivided into three categories:

- Rough grids: Where the tolerance of the linear sizes is more than 100 microns and angular one are more than 5 minutes;
- Normal grids: Where the tolerance of the linear sizes is in the range from 10 to 100 um and angular one are in the 1 - 5 minutes range;
- Precise grids: Where the tolerance of the linear sizes is in the range from 1 to 10 um and angular one are in the 1 - 60 seconds range.

Conventionally, there are two fundamentally different application methods of grids making, which are chosen depending on the configuration of the line and the required class of accuracy. One of that is the method of division, and other photolithography method.

The method of division includes two types:

1) Line making by a diamond cutter directly on the glass surface (mechanical method). It is used in the manufacture of grids with width from 0.5 up to 10 um.

2) Grid making is by scribing of protective layer that cover the glass surface with its following etching (glazing of the protective layer). This method allows obtaining of lines with a 10 - 20 um.

Photolithography allows get a complex drawings with high accuracy of size, but that is technologically complicated. The minimum width of the lines, which can be obtained by this way, is 3 um [1].

Recently laser controlled thermocracking (LCT) method has got a wide application for the precise cutting of brittle non-metallic materials [2,3].

THE USE OF THE LCT

The formation of separating cracks occurs due to the tension stress arising as a result of a material surface heating by a laser radiation with following cooling of the heating zone by a coolant. The crack is formed in the material because of the impact of tension stresses where the cutting depth can be tuned from several microns up to through cut. Using air-water aerosol as a coolant during a cutting process, this crack can be visually observed in about the 3 - 10 mm distance after the water jet stream (**Figure 1**), from the place of that forming. Then the microcrack is closed because of stresses arising in the volume of material in the moment of microcrack formation. The microcrack is fully interlocked because of the edge of one does not have any defects and that becomes visually invisible. Therefore, the use of such microcracks as optical lines and grids is not possible. In addition, when cutting of materials by microcrack (pre-cut) it creates difficulties in following braking of the workpiece into the separate parts which often leads to the yield decreasing.

The task of lines and grids making for optical instruments has been successfully solved using LCT method for the quartz glass [4]. In this case, due to the presence of residual thermostress along the cutting line, the microcrack closing is not happened. Varying the relative parameters of the LCT process, namely the feed rate of relative movement and the density of the laser radiation power, it is possible to control the microcrack with strictly defined geometrical parameters as width and depth. The tunable range of these parameters is very wide. For instance, when reducing the feed rate of relative movement during LCT process from 50 mm/sec down to 10 mm/sec, the width of microcracks is practically linearly increases from 0.5 to 8 um (**Figure 2**).

During R & D of the LCT method with using of different mixed compounds for the coolant the fundamentally new application method has been developed for grids and scales making for surfaces of optical parts [5]. This method carries the formation of microcrack in the glass surface with the width and depth, and it also carries filling the microcrack by various compounds for its visualization, and it also carries the control of optical and geometric parameters of this microcrack.

To obtain a visible cutting line (microcrack) with the specified width and depth it is necessary to use a coolant with the two-level dispersion consisting from a dispersion air environment and the two-phase compound of disperse phase. Thereby the first dispersed phase includes water drops and the second disperse phase includes the colloid compound or solid microparticles.

In addition, for microcracks visualization, cooling of the cutting line can be realized with the coolant in the form of a disperse system containing particles of different colors in the dispersed phase or with the injection in the composition of the disperse phase using disperse coloring agents with different colors or colored solutions.

The line making process by LCT method is as follows (**Figure 1**). When local heating of glass workpiece 1 by laser elliptical beam 2 is focused through spherical and cylindrical lenses 3 compression stresses is created in the material surface layer, which do not exceed the strength of the material and does not lead to its damage. Following sharp cooling of the heating zone by the coolant 5 sprayed into a heating zone by nozzle 4 leads to transforming of compression stresses to tensile stresses, which exceed the strength of the glass and lead microcrack creating in the glass surface 6. Solid particles of the dispersed phase that are in the composition of the coolant put in the microcrack when nozzle spraying and that particles are fixed inside microcrack then preventing its collapse (closing or growing together). For instance, the special dispersed suspension was used as a coolant, where the dispersive environment was the airwater mixture and the disperse phase was a water dispersion of silica where the average SiO_2 particle size was about 10 nm. When you move the sample glass with a 250 mm/s feed rate, the laser beam is pre-cut with 120 microns microcrack depth, thus the width of one was 2 microns.

Using phosphor solutions in the disperse phase, the microcrack will be illuminated as a glowing line (**Figure 3**(a)) if to backlight transversely and it will be dark black line (**Figure 3**(b)) if to backlight in longitudinal manner. Crack depths were gotten by LCT method with a range from tens to several hundreds of micrometers. The width of that may be varied from tens of nanometers (when using concentrated colloidal solutions) and to tens of micrometers (when using solid fine-dispersed powders).

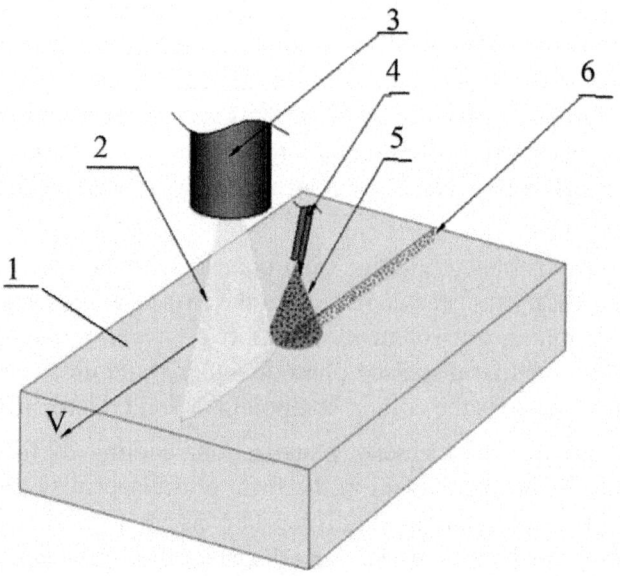

Figure 1: Scheme is the microcrack formation in the glass: 1—A glass workpiece; 2—The laser beam; 3—Focusing lens; 4—Water jet; 5—Coolant; 6—Microcrack.

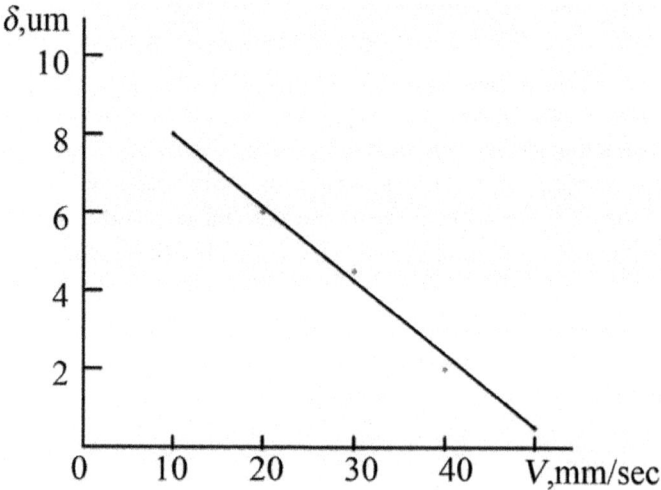

Figure 2: The dependence of the line's (microcrack) width d from the feed rate V when LCT method.

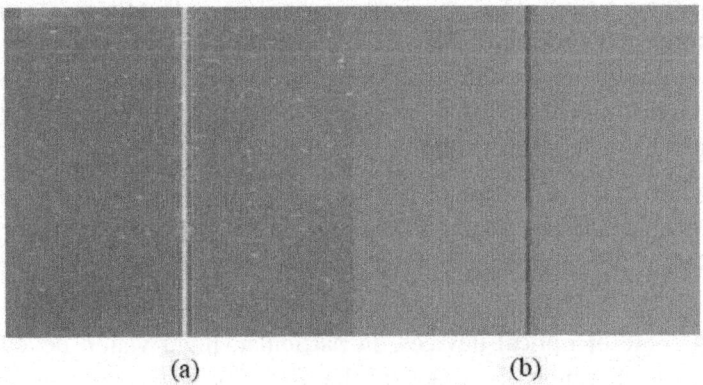

(a)   (b)

Figure 3: Microcrack view is with backlight (a) and without backlight (b).

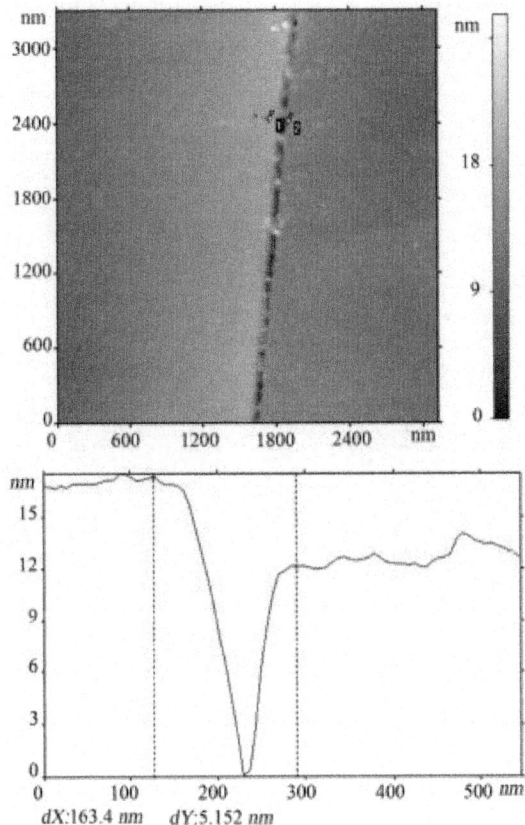

dX:163.4 nm   dY:5.152 nm

Figure 4: shows results of microcrack geometrical parameters measurements filled by a phosphor that was measured on the tunnel microscope Supra 50 VP LEO.

The main advantage of developed technology in that the production of grid lines is the single technological cycle, i.e. following etching and line painting is not required (as for the mechanical method) and there are no complex technological processes (as for the photolithography). The speed of the line making is a few hundred millimeters per second.

An important advantage of this method is that practically any material can be put into the microcrack (in the form of fine particles) with required optical properties, without worrying about the presence or absence of adhesion to a glass.

For some of optical devices, in particular, night vision devices, one of the main requirements is to ensure the visibility of the grid lines on a dark background. The part of light is scattered on the glass surface and smallest defects (if it there are) at the common backlight. It gives a common light background that confuses the outlook. In these cases an effective solution is luminescent lines making, which lets be highlighted by ultraviolet light that is invisible for the human eye (**Figure 5**(a)).

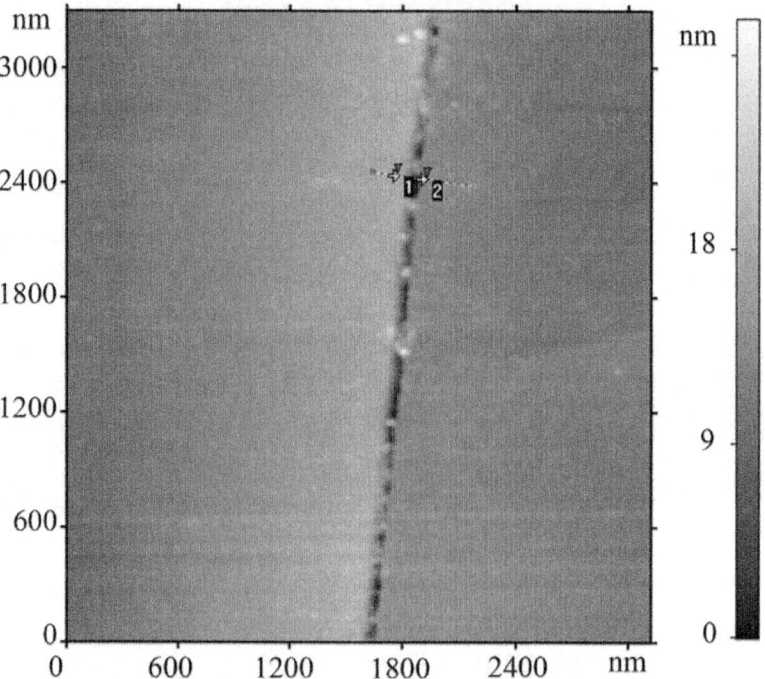

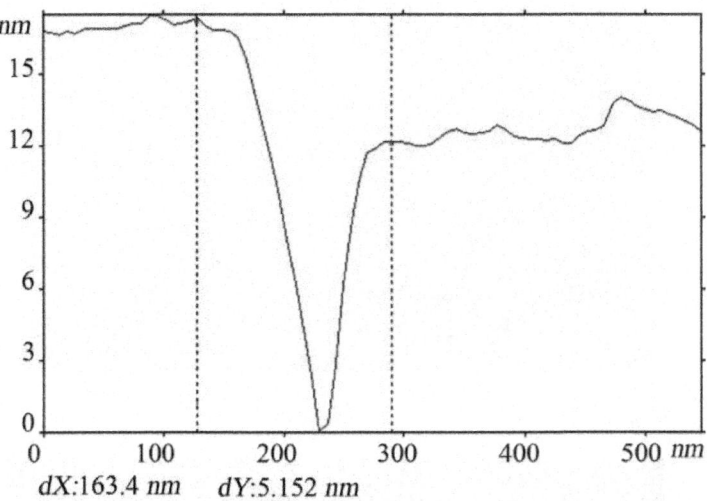

dX:163.4 nm   dY:5.152 nm

Figure 4: Results are measurements of geometrical parameters of microcrack filled by a phosphor. Measuring equipment is the tunnel microscope Supra 50 VP LEO.

Manufacturing of such lines by traditional methods is difficult because not all of phosphors have a good adhesion to a glass. Using the LCT method, this kind of the problem doesn't arise. Various phosphors (including organic) can be successfully used with the possibility of obtaining different lines colors and backlight in the different spectral range (**Figure 5**(b)). Varying by phosphor concentration in the coolant, it gives lines with a different intensity of the glow. The width of the grid lines is only a few hundredths of a micron but they are clearly visible as clear luminous lines, both as in case of magnification and with the naked eye.

It should be noted that the described method for optical line application is not limited for continuous lines. **Figure 6** shows photos of the dash line (a) and dash quintuple-dot line (b) that was obtained by the LCT method. It is available to make many kinds of linetypes depending on the request.

CONCLUSIONS

The new high-effective method allows to obtain optical tags: Line, mark, risk, grid, etc. with the set of geometrical parameters that impossible to get using conventional technologies.

In addition, the described method [5] provides the possibility to make wiring in the glass (for instance) filling the microcrack by metal current-conducting nanopowders and it provides to get glowing grooves using luminescent compounds.

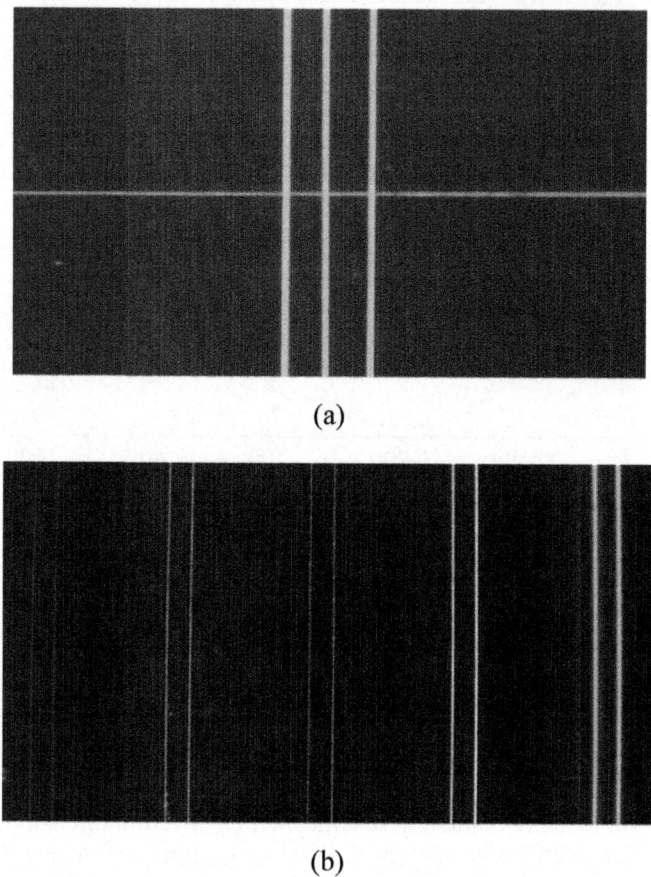

(a)

(b)

Figure 5: Luminous lines of blue (a) and different colors (b).

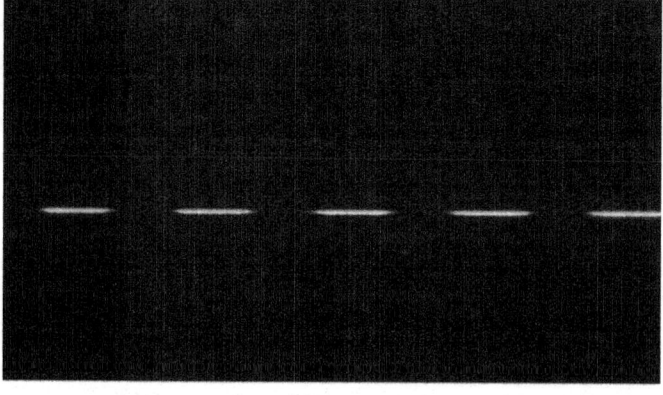

(a)

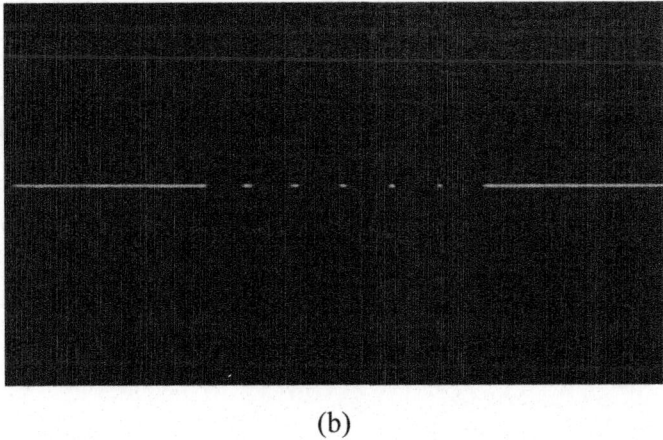

(b)

Figure 6: The dash line (a) and dash quintuple-dot line (b) photos obtained by the LCT method.

It can be widely used in the building market, for some new generation devices and products.

The fulfilled complex of works in fundamental and applied research of the LCT method shows the competitiveness of the Russian scientific school in the field of high technologies.

Many researches in this paper were done in accordance with the Government Decree of the Russian Federation No. 218, dated 9 April, 2010, on the basis of a Contract 13. G25. 31.0020 between the Ministry of Education and Science of the Russian Federation and JSC "MZ 'Sapphire'" in creation of hi-tech production manufacturing.

REFERENCES

1. L. M. Krivoviaz, D. T. Puriaev and M. A. Znamenskaya, "The Practice of Optical Measuring Laboratory," Machinery, Moscow, 2004, 333 p.

2. V. S. Kondratenko, "Method of Cutting of Brittle Material," the Patent of Russian Federation, No. 2024441, 1991.

3. V. Golubjatnikov, V. S. Kondratenko and A. B. Zhymalov, "Development of the Theory and Practice of a Method Laser Controlled Thermocracking," Instruments, Vol. 12, No. 114, 2012, pp. 1-6.

4. V. S. Kondratenko, V. I. Gundiak and V. Y. Bersenyev, "Application of the Method of Laser Controlled Thermocracking of a Glass in the Production of Optical Parts— Electronic Engineering, Ser," Laser Technology and

Optoelectronics, Ser. Electronic Engineering, Vol. 4, No. 56, 1990, pp. 70-71.

5. V. S. Kondratenko, "Method of Cutting of Brittle NonMetallic Materials (Two Variants)," The Patent of Russian Federation, No. 2333163, 2007.

Chapter 6

ADDITIVE MANUFACTURING OF CO-CR-MO ALLOY: INFLUENCE OF HEAT TREATMENT ON MICROSTRUCTURE, TRIBOLOGICAL, AND ELECTROCHEMICAL PROPERTIES

Kedar Mallik Mantrala[1], Mituns Das[2], Krishna Balla[2], Ch. Srinivasa Rao[3] and V. V. S. Kesava Rao[3]

[1]Department of Mechanical Engineering, Vasireddy Venkatadri Institute of Technology, Guntur, India

[2]Bioceramics and Coating Division, CSIR-Central Glass and Ceramic Research Institute (CGCRI), Kolkata, India

[3]Department of Mechanical Engineering, Andhra University College of Engineering, Visakhapatnam, India

ABSTRACT

Co-Cr-Mo alloy samples, fabricated using Laser Engineered Net Shaping – a laser-based additive manufacturing technology, have been subjected to heat treatment to study its influence on microstructure, wear, and corrosion properties. Following L9 Orthogonal array of Taguchi method, the samples were solutionized at 1200°C for 30, 45, and 60 min followed by water quenching. Aging treatment was done at 815 and 830°C for 2, 4, and 6 h. Heat treated samples were evaluated for their microstructure, hardness, wear resistance, and corrosion resistance. The results revealed that highest hardness of 512 ± 58 Hv and wear rate of $0.90 \pm 0.14 \times 10^{-4}$ mm^3/N·m can be achieved with appropriate post-fabrication heat treatment. Analysis of variance and gray relational analysis on the experimental data revealed that the samples subjected to solution treatment for 60 min, without aging, exhibit best combination of hardness, wear, and corrosion resistance.

INTRODUCTION

Co-Cr-Mo alloys are well known for their high wear resistance and corrosion resistance. As they also exhibit good biocompatibility, they are widely used as surgical prostheses (Krishna et al., 2008; Bandyopadhyay et al., 2009; España et al., 2010; Dittrick et al., 2011). The microstructure and properties like wear resistance, hardness, etc., of Co-Cr-Mo alloy depends mainly on the carbon content and the type of heat treatment applied. Many researchers have applied different heat treatments to understand the mechanical behavior of Co-Cr-Mo alloys. In many heat treatment experiments, $M_{23}C_6$ type carbides were observed to precipitate along the grain boundaries in the γ matrix of the alloy (Giacchi et al., 2011). In Ni and C free Co-Cr-Mo alloys, high heat treatment temperatures found to suppress γ (fcc) to ε (hcp) martensitic transformation (Lee et al., 2005). Further, morphological change from lamellar to round carbides has been reported during solution and partial solution treatments of this alloy. Optimal results were obtained with solutionizing at 1120°C for 1 h followed by an aging at 815°C for 4 h (Bedolla-Gil et al., 2009).

As cast ASTM F75 – Co-Cr-Mo alloy usually exhibit two phase microstructure consisting of $\sigma/M_{23}C_6$ and σ/Co-α eutectic (Rosenthal et al., 2010). The π phase has been identified in these alloys with carbon content between 0.15 and 0.35% after heat treatment at 1275°C and holding times up to 0.5 h. Further complete dissolution of precipitates was observed in alloys with carbon contents <0.35% (Mineta et al., 2010). When Co-Cr-Mo cast alloy samples were subjected to a solution heat treatment at 1230°C for 3 h followed by water quenching and isothermal aging at 850°C for different times, the samples exhibited lamellar type carbides (γ FCC + $M_{23}C_6$) along the grain boundaries (Lashgari et al., 2011). Solution heat treatment carried out at 1225°C for 240 min on Co-Cr-Mo alloy resulted in blocky $M_{23}C_6$ carbides in lamellar type matrix structures (Giacchi et al., 2012). Moreover, the microstructure of hot forged Co-Cr-Mo was found to be finer and exhibited superior mechanical properties compared to annealed alloys (Okazaki, 2008). Some reports are available on tribo-corrosion behavior of Co-Cr-Mo alloys (Ortega-Saenz et al., 2011;Pourzal et al., 2011; Mischler and Munoz, 2013). From these earlier studies, it is clear that no reported literature is available on the influence of heat treatment on microstructure, tribological, and corrosion properties of laser processed Co-Cr-Mo alloy.

Our earlier study on laser fabrication of Co-Cr-Mo alloy revealed that metallurgically sound, dense components can be obtained with a fine microstructure, when the process parameters are optimized (Mantrala et al., 2014). When compared with wrought Co-Cr-Mo alloy, the carbide volume fraction and the hardness were comparable, but the abrasive wear resistance of laser deposited alloy was found to be less (Janaki Ram et al., 2008). However, our preliminary studies on laser assisted additive manufacturing of Co-Cr-Mo alloy demonstrated that high laser power, low powder feed rate, and high scan speed can produce deposits with excellent mechanical, tribological, and electrochemical properties (Mantrala et al., 2014). In the present investigation, we aim to understand the influence of post-fabrication heat treatment on microstructure, hardness, wear, and corrosion properties of laser deposited Co-Cr-Mo alloy.

MATERIALS AND METHODS

Commercially available Co-Cr-Mo alloy (Stellite 21 – Kennametal Stellite, Goshen, IN, USA) powder was used to fabricate Co-Cr-Mo alloy using Laser Engineered Net Shaping (LENS™). Several deposits consisting of 5–8 layers and 15 mm² were fabricated at 350 W laser power, 5 g/min feed rate, and 20 mm/s scan speed to study the influence of post-fabrication heat treatment on microstructure, wear, and corrosion properties of this alloy. The above laser process parameters were optimized in our earlier study (Mantrala et al., 2014) and typical component fabricated using these parameters is shown in Figure 1.

Figure 1

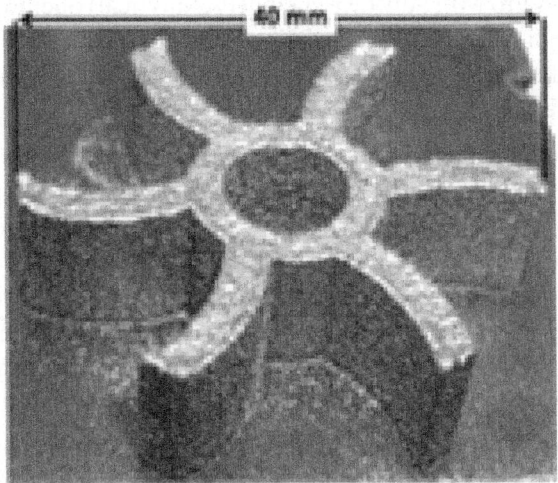

Figure 1: Additively manufactured Co-Cr-Mo alloy impeller using Laser Engineered Net Shaping (LENS™).

L9 Orthogonal array of Taguchi method has been applied to optimize heat treatment cycle (Table1). The Taguchi method provides minimum number of balanced experiments with small variance of results. All the samples were water quenched after solution heat treatment at 1200°C. The samples S1, S4, and S7 were restricted to solution treatment only whereas the other samples were subjected to aging at 815 and 830°C for different timings shown in Table 1.

Table 1: Experimental parameters used in the present investigation

Sample no.	Solution treatment time (min)	Aging temperature (°C)	Aging time (h)
S1	30	No aging	
S2	30	815	4
S3	30	830	6
S4	45	No aging	
S5	45	815	6
S6	45	830	2
S7	60	No aging	
S8	60	815	2
S9	60	830	4

The deposit hardness was measured using Vickers microhardness tester (MMT X7 MATSUZAWA) at 200 g load applied for 20 s. An average of 10 measurements on 3 identical samples was reported. The constituent microstructural phases of laser deposited Co-Cr-Mo alloy were determined using an X'Pert Pro MPD diffractometer (PANalytical) operating at 45 kV and 40 mA using Ni-filtered CuKα radiation. The microstructures of the samples were analyzed using scanning electron microscope (ProX, Phenom-World BV, Eindhoven, Netherlands).

Electrochemical measurements were performed on heat treated Co-Cr-Mo alloy samples using a potentiostat/galvanostat (SP300, Bio-Logic SAS, France) in 3.5% NaCl solution. Before testing, the top surfaces of the deposits were polished up to 1 μm Al_2O_3. Potentiodynamic polarization tests were performed (on three identical samples) by scanning the applied potential from −0.5 to 1.6 V vs. saturated calomel electrode (SCE) at a rate of 10 mV/min. The corrosion potential (E_{corr}) and the corrosion current density (I_{corr}) was extracted from the polarization curves of the linear part of the cathodic curve (Tafel behavior).

Ball on disk wear tests were performed on the samples using a tribometer (NANOVEA, Microphotonics Inc., CA, USA) under a normal load of 10 N for 1000 m sliding distance. An Al_2O_3 ball of 3 mm diameter was used as counter body. The wear tests were carried out on three samples processed at each heat treatment condition (S1–S9). The wear rate of the deposits was calculated from the wear track measurements and was represented as $mm^3/N \cdot m$. The wear tracks were observed using SEM to understand the wear mechanism.

Analysis of variance (ANOVA) was performed on the hardness, wear rate, and corrosion data. In the present work, we aim at identifying optimum heat treatment parameters, which can provide high corrosion resistance along with high hardness and wear resistance. For this purpose, the multiple responses (hardness, wear rate, E_{corr}, and low I_{corr}) have been converted in to a single index using gray relational grade analysis (Deng, 1989). It is known that a material with more positive E_{corr} and low I_{corr} exhibit high corrosion resistance. Therefore, we have used "higher the better" for E_{corr} and hardness, and "lower the better" for I_{corr} and wear rate for normalization. Following equations were used to calculate the gray relational grade (Deng, 1989).

Normalization of Response

$$\text{Higher the better} = N_i(R) = \frac{\nu_i(R) - \nu_i(R)(\min)}{\nu_i(R)(\max) - \nu_i(R)(\min)}$$

(1)

$$\text{Lower the better} = N_i(R) = \frac{\nu_i(R)(\max) - \nu_i(R)}{\nu_i(R)(\max) - \nu_i(R)(\min)}$$

(2)

where $N_i(R)$ is the normalized response, $v_i(R)$ is the experimental value of Rth response (in the present work the responses were hardness, wear rate, E_{corr} and I_{corr}), $v_i(R)(\max)$ and $v_i(R)(\min)$ are maximum and minimum experimental values of Rth response.

Gray Relational Coefficient

The normalized response $[N_i(R)]$ was used to calculate gray relational coefficient as follows:

$$\alpha(R) = \frac{\Delta_{\min} + 0.5\Delta_{\max}}{\Delta v_i(R) + 0.5\Delta_{\max}}$$

(3)

where $\Delta v_i(R) = 1 - N_i(R)$, $\Delta_{\min}$ and $\Delta_{\max}$ are the minimum and maximum values of $\Delta v_i(R)$ (for Rth response).

Gray Relational Grade

The final grading of combined effect of responses was done with gray relational grade as follows:

$$\text{GRG} = \frac{1}{n} \sum_{R=i}^{n} \propto (R)$$

(4)

where n is the number of responses (in the present work the responses were hardness, wear rate, E_{corr}, and I_{corr}, i.e., n = 4). The highest grade indicates the best combination of process parameters for high hardness and E_{corr}, and low wear rate and I_{corr}.

RESULTS AND DISCUSSION

The microstructures of laser deposited Co-Cr-Mo alloy heat treatment (S1–S9) are shown in Figure2. In general, all samples showed equiaxed Co grains with or without carbide precipitates depending on heat treatment conditions. However, heat treatments found have no measurable influence on grain size of laser fabricated Co-Cr-Mo alloy. The carbides precipitates were observed not only along the boundaries but also within the grains.

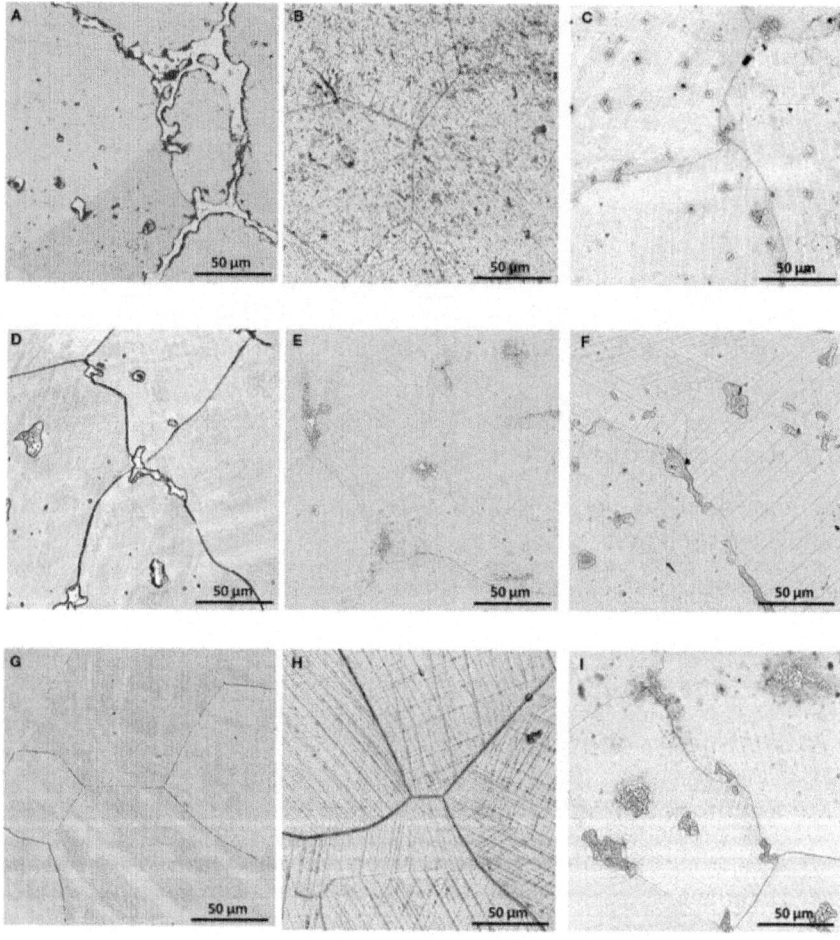

Figure 2: Microstructure of Co-Cr-Mo alloy samples solutionized at 1200° C for different times followed by aging (A) S1-30 min, no aging; (B) S2-30 min, 815°C, 4 h; (C) S3-30 min, 830°C, 6 h; (D) S4-45 min, no aging; (E) S5-45 min, 815°C, 6 h; (F) S6-45 min, 830°C, 2 h; (G) S7-60 min, no aging; (H) S8-60 min, 815°C, 2 h; (I) S9-60 min, 830°C, 4 h.

The precipitates found to decrease in size and amount with increasing solutionizing time from 30 to 60 min, as shown in Figures 2A,D,G. At solutionizing temperature of 1200°C, increasing the holding time allow more diffusion and carbides can dissolve in the matrix and therefore the concentration of precipitates decreased in Co-Cr-Mo alloy with increasing solutionizing time. As expected, increasing the aging time from 2 h (Figure2F) to 4 h (Figure 2I), at 830°C aging temperature, increased the precipitation and is attributed to decrease in solid solubility of carbide elements in Co matrix.

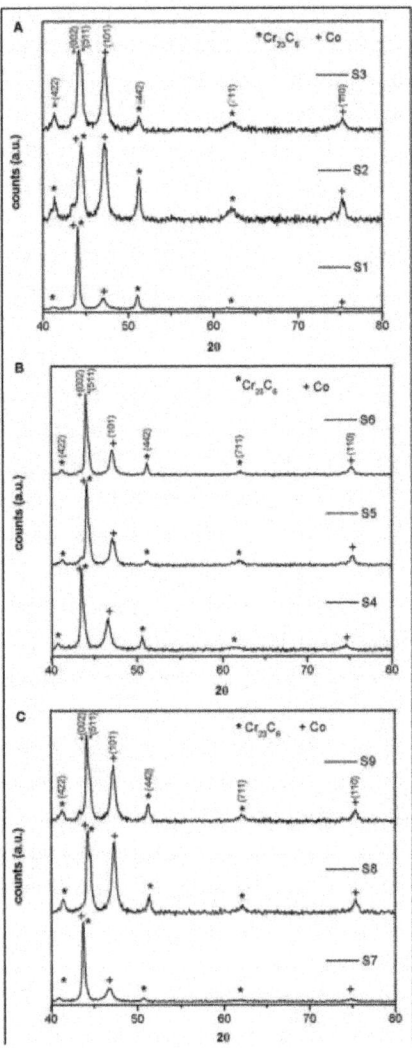

Figure 3: Influence of heat treatment on constituent phases in laser processed Co-Cr-Mo alloy. (A) S1 to S3, (B) S4 to S6, (C) S7 to S9.

Similarly, the increasing the aging temperature increased the precipitation concentration as shown in Figure 2C (S3-30 min, 830°C, 6 h) and Figure 2E (S5-45 min, 815°C, 6 h). X-ray diffraction data, Figure 3, confirmed the presence of carbide precipitates in most of Co-Cr-Mo alloy samples. The precipitates were identified as $Cr_{23}C_6$ precipitates and the aging treatment increased their peak intensities. This observation indicates that the precipitation increased with increasing aging time and temperature. The XRD results also confirmed the absence of any other type of carbides in laser fabricated Co-Cr-Mo alloy.

The influence of heat treatment on hardness of Co-Cr-Mo alloy is presented in Table 2. Although the microstructures and XRD showed some clear trend with solutionizing time, aging temperature, and time, mixed trend was observed for hardness. Some samples (S2, S3, S4, S5, and S9) showed low hardness between 372 and 386 Hv, whereas other samples (S1, S6, S7, and S8) exhibited high hardness in the range of 478–512 Hv. Highest hardness of 512 ± 58 Hv was recorded for samples, which were solutionized for 45 min at 1200°C followed by aging at 830°C for 2 h. However, tribological testing showed lowest wear rate (0.90 ± 0.14 × 10^{-4} mm³/N·m) with heat treatment cycle of 60 min at 1200°C followed by aging at 830°C, 4 h. While the sample, which exhibited high hardness of 491 ± 51 Hv found to have high wear rate of 1.44 ± 0.28 × 10^{-4} mm³/N·m during tribological testing. This discrepancy is presumably due to variations in precipitate size and their distribution as a result of post-fabrication heat treatment. In general, the samples subjected to longer solutionizing times (S7, S8, S9) exhibited high wear resistance compared to other samples, which could be due to complete dissolution and re-precipitation of carbides during solutionizing and aging treatments, respectively. The wear track morphologies of Co-Cr-Mo alloy samples are presented in Figure 4. All samples showed deep grooves with third body wear characteristics. The samples with low wear rate exhibited smoother wear tracks than other samples with high wear rate. For example, the sample S4 (solutionized for 45 min at 1200°C) and S9 (solutionized for 60 min at 1200°C and aged 830°C for 4 h) showed clearly smoother wear tracks (Figures 4D,I) compared to other samples. Further, in all samples carbide pullout was observed as shown in insets of Figures 4D,G. These pulled out carbide particles entrapped between articulating substrate and Al_2O_3 ball resulted in third body wear.

Table 2: Experimentally determined hardness, wear rate, and corrosion data of laser fabricated Co-Cr-Mo alloy

Sample ID	Hardness (Hv)	Wear rate ($\times 10^{-4}$ mm^3/N·m)	Corrosion		Gray relational analysis	
			I_{corr} (μA/cm^2)	E_{corr} (mV vs. SCE)	Grade	Order
S1	491 ± 51	1.44 ± 0.26	0.11	−368	0.563	7
S2	382 ± 25	1.18 ± 0.24	0.20	−300	0.463	9
S3	378 ± 11	1.19 ± 0.25	0.22	−61	0.514	8
S4	372 ± 44	0.96 ± 0.28	0.42	+64	0.624	3
S5	386 ± 25	1.17 ± 0.12	0.44	+85	0.568	6
S6	512 ± 58	1.08 ± 0.18	0.28	−116	0.670	2
S7	478 ± 81	0.96 ± 0.25	0.05	−178	0.739	1
S8	484 ± 37	0.97 ± 0.13	0.61	−86	0.603	5
S9	380 ± 35	0.90 ± 0.14	0.39	−22	0.619	4

Gray relational grade and its order are presented in the last column.

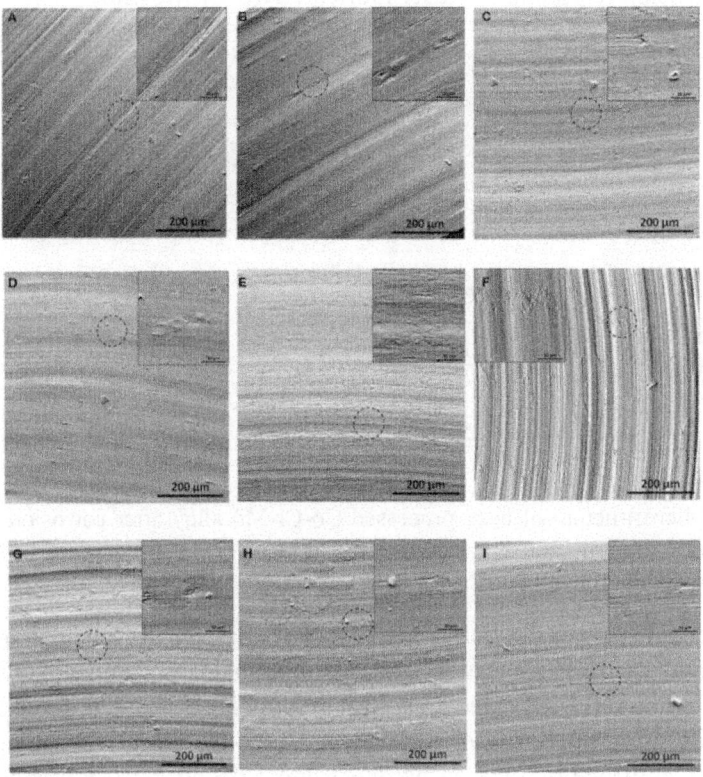

Figure 4: Wear track morphologies of Co-Cr-Mo alloy samples (A–I) S1–S9. Inset shows magnified image of region highlighted with circle.

The corrosion potential (E_{corr}) and corrosion current (I_{corr}) of Co-Cr-Mo alloy samples derived from respective Tafel plots are summarized in Table 2. It is well known that the materials exhibit high corrosion resistance if E_{corr} values tend to be positive with minimum possible I_{corr} values. Our experimental results show that I_{corr} values decreased from 4.63 $\mu A/cm^2$ (un-heat treated samples) (Mantrala et al., 2014) to the range of 0.05–0.61 $\mu A/cm^2$ after heat treatment, while mixed results were observed with E_{corr} values. Evidently, the corrosion rate is relatively lower than untreated samples (Mantrala et al., 2014). As shown in Figure 5, the corrosion was found to be localized primarily along grain boundaries. In addition, the corrosion was more pronounced around the carbide precipitates. This is primarily due to reduction in Cr in the matrix as a result of carbide precipitation. To study the combined effect of I_{corr} and E_{corr}, Gray Relational Analysis has been carried out and the results revealed that samples solutionized at 1200°C for 45 min without aging would provide highest corrosion resistance for laser fabricated Co-Cr-Mo alloy.

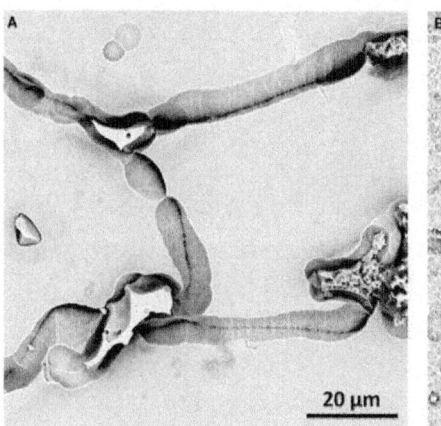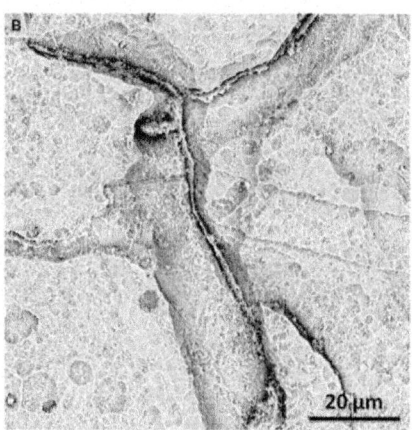

Figure 5: Microstructure of laser processed Co-Cr-Mo alloy after corrosion testing (A) S7, solutionized at 1200°C for 60 min, no aging; (B) S9, solutionized at 1200°C for 60 min, aged for 4 h at 830°C.

The percentage contribution of heat treatment parameters toward hardness, wear rate, and corrosion resistances was calculated using ANOVA and tabulated in Table 3. The data indicated that heat treatment parameters have strong influence on properties of laser fabricated Co-Cr-Mo alloy. The aging time had highest influence of 77% on hardness and 57% on corrosion resistance. Whereas the most influential factor for wear resistance was solutionizing time with 71%. For example, samples those were not subjected to aging (or aged for shorter times) exhibited high hardness and high corrosion resistance (S1, S6, S7, and S8). Similarly, irrespective of aging time, the samples subjected to longer solution times have exhibited high wear resistance (S7, S8, and S9). Based on above discussion for high hardness, high wear resistance, and high corrosion resistance samples S6, S9, and S7 are the best, respectively. To identify the best heat treatment parameters for best combination of high hardness, high wear resistance, and high corrosion resistance the gray relational analysis was used. The analysis showed that laser fabricated Co-Cr-Mo alloy can provide best combination of hardness, wear, and corrosion resistance if a solution treatment at 1200°C for 60 min is given. In summary, the solutionizing time found to have stronger influence followed by aging temperature and time on overall performance of the Co-Cr-Mo alloy.

Table 3: Influence of heat treatment parameters on hardness, wear, and corrosion resistance

Response	Solution time	Aging temperature	Aging time
Hardness (%)	5	5	77
Wear resistance (%)	71	2	14
Corrosion resistance (%)	13	27	57
Overall (%)	56	29	5

CONCLUSIONS

In this work, we have successfully fabricated net shape Co-Cr-Mo alloy impeller using laser based additive manufacturing technology. Microstructural analysis showed complete carbide dissolution with increasing solutionizing time and their precipitation with aging temperature and time. Overall, the heat treatment found to improve properties of laser fabricated Co-Cr-Mo alloy. The samples solutionized at 1200°C for 45 min followed by aging at 830°C for 2 h exhibited highest hardness. Tribological tests on these samples revealed that solutionizing treatment at 1200°C for 60 min with 4 h aging at 830°C results in significant improvement in wear resistance of this alloy.

CONFLICT OF INTEREST STATEMENT

The authors declare that the research was conducted in the absence of any commercial or financial relationships that could be construed as a potential conflict of interest.

ACKNOWLEDGMENTS

The authors would like to acknowledge the financial support from the Council of Scientific and Industrial Research (CSIR), New Delhi to establish LENS™ facility at CSIR-CGCRI and through project ESC0103. Mr. KM Mantrala is grateful to the Director, CSIR-CGCRI, Kolkata for granting permission to carry out some of the experimental and characterization work at CSIR-CGCRI.

REFERENCES

1. Bandyopadhyay, A., Krishna, B. V., Xue, W., and Bose, S. (2009). Application of laser engineered net shaping (LENS) to manufacture porous and functionally graded structures for load bearing implants. J. Mater. Sci. Mater. Med. 20(Suppl. 1), S29–S34. doi:10.1007/s10856-008-3478-2

2. Bedolla-Gil, Y., Juarez-Hernandez, A., Perez-Unzueta, A., Garcia-Sanchez, E., Mercado-Solis, R., and Hernandez-Rodriguez, M. A. L. (2009). Influence of heat treatments on mechanical properties of a biocompatility alloy ASTM F75. Rev. Mex. Fis. 55, 1–5.

3. Deng, J. (1989). Introduction to grey system. J. Grey Syst. 1, 1–24.

4. Dittrick, S., Balla, V. K., Davies, N. M., Bose, S., and Bandyopadhyay, A. (2011). In vitro wear rate and Co Ion release of compositionally and structurally graded CoCrMo-Ti6Al4V structures. Mater. Sci. Eng. C 31, 809–814. doi:10.1016/j.msec.2010.07.009

5. España, F. A., Balla, V. K., Bose, S., and Bandyopadhyay, A. (2010). Design and fabrication of CoCrMo based novel structures for load bearing implants using laser engineered net shaping. Mater. Sci. Eng. C 30, 50–57. doi:10.1016/j.msec.2009.08.006

6. Giacchi, J. V., Fornaro, O., and Palacio, H. (2012). Microstructural evolution during solution treatment of Co-Cr-Mo-C biocompatible alloys. Mater. Charact. 68, 49–57. doi:10.1016/j.matchar.2012.03.006

7. Giacchi, J. V., Morando, C. N., Fornaro, O., and Palacio, H. A. (2011). Microstructural characterization of as-cast biocompatible Co-Cr-Mo alloys. Mater. Charact. 62, 53–61. doi:10.1016/j.matchar.2010.10.011

8. Janaki Ram, G. D., Esplin, C. K., and Stucker, B. E. (2008). Microstructure and wear properties of LENS deposited medical grade CoCrMo. J. Mater. Sci. Mater. Med. 19, 2105–2111. doi:10.1007/s10856-007-3078-6

9. Krishna, B. V., Xue, W., Bose, S., and Bandyopadhyay, A. (2008). Functionally graded Co-Cr-Mo coating on Ti-6Al-4V alloy structures. Acta Biomater. 4, 697–706. doi:10.1016/j.actbio.2007.10.005

10. Lashgari, H. R., Zangeneh, S. H., and Ketabchi, M. (2011). Isothermal aging effect on the microstructure and dry sliding wear behavior of Co-28Cr-5Mo-0.3C alloy. J. Mater. Sci. 46, 7262–7274. doi:10.1007/s10853-011-5686-2

11. Lee, S.-H., Takahashi, E., Nomura, N., and Chiba, A. (2005). Effect of heat treatment on microstructure and mechanical properties of Ni- and C-free Co-Cr-Mo alloys for medical applications. Mater. Trans. 46, 1790–1793. doi:10.2320/matertrans.46.1790

12. Mantrala, K. M., Das, M., Balla, V. K., Rao, Ch. S., and Kesava Rao, V. V. S. (2014). Laser-deposited CoCrMo alloy: microstructure, wear and electrochemical properties. J. Mater. Res. 29, 2011–2027. doi:10.1557/jmr.2014.163

13. Mineta, S., Namba, S., Yoneda, T., Ueda, K., and Narushima, T. (2010). Carbide formation and dissolution in biomedical Co-Cr-Mo alloys with different carbon contents during solution treatment. Metall. Mater. Trans. A. 41A, 2129–2138. doi:10.1007/s11661-010-0227-1

14. Mischler, S., and Munoz, A. I. (2013). Wear of CoCrMo alloys used in metal-on-metal hip joints: a tribocorrosion appraisal.Wear 297, 1081–1094. doi:10.1016/j.wear.2012.11.061

15. Okazaki, Y. (2008). Effects of heat treatment and hot forging on microstructure and mechanical properties of Co-Cr-Mo alloy for surgical implants. Mater. Trans. 49, 817–823. doi:10.2320/matertrans.MRA2007299

16. Ortega-Saenz, J. A., Hernandez-Rodriguez, M. A. L., Ventura-Sobrevilla, V., Michalczewski, R., Smolik, J., and Szczerek, M. (2011). Tribological and corrosion testing of surface engineered surgical grade CoCrMo alloy. Wear 271, 2125–2131. doi:10.1016/j.wear.2010.12.062

17. Pourzal, R., Catelas, I., Theissmann, R., Kaddick, C., and Fischer, A. (2011). Characterization of wear particles generated from CoCrMo alloy under sliding wear conditions. Wear 271, 1658–1666. doi:10.1016/j.wear.2010.12.045

18. Rosenthal, R., Cardoso, B. R., Bott, I. S., Paranhos, R. P. R., and Carvalho, E. A. (2010). Phase characterization in as-cast F-75 Co-Cr-Mo-C alloy. J. Mater. Sci. 45, 4021–4028. doi:10.1007/s10853-010-4480-x

Chapter 7

ANALYSIS OF TECHNOLOGY EFFECTIVENESS OF LEAN MANUFACTURING USING SYSTEM DYNAMICS

Hasan Hosseini-Nasab, Mohammad Dehghani, and Amin Hosseini-Nasab

School of Engineering, Yazd University, P.O. Box 89195-741, Yazd, Iran

ABSTRACT

In today's competitive environment, organizations are seeking to improve their position in the market. Lean manufacturing is an effective tool for elevating the competitiveness of organizations based on the fact that each can find its own way of improvement. Technology improvement is considered to be one of lean manufacturing's dimensions. Technology is defined as the usage and knowledge of tools, techniques, crafts, systems, or methods of organization, with the aim of solving a problem or creating an artistic perspective. A dynamic model could be appropriate for analyzing the interrelated behavior of technology and lean manufacturing. Despite the fact that there are plenty of papers and case studies on the applications of Lean manufacturing in organizations, only a few are focused on the dynamic aspects of the system. In this paper, a dynamic model is presented in which Lean manufacturing is linked with technology by causal relationships. The notable advantage of the presented model is the ability to alter a parameter to find how it affects others parameters by considering key results. Thus, it is reasonable to expect that the results of such analysis could somewhat improve the efficiency of technology improvement on Lean manufacturing.

INTRODUCTION

Lean philosophy has been prevalent in the manufacturing industry during recent decades. However, even today, a large number of manufacturers are struggling to successfully embrace Lean principles. Results of the Aberdeen Group's report on lean scheduling indicate that, among manufacturers, those who adopted Lean software applications have exceeded many of their competitors, thus making Lean software adoption vital for manufacturers. What manufacturers require is to standardize Lean processes across their enterprise by establishing a Lean center and providing factory floor data as actionable intelligence, in order to successfully leverage their investments in technology [1].

In today's competitive world with its vast and rapid changes in scientific-technical areas and continuous challenges in economical-social systems, there are still many firms with a suitable position. These firms are flexible, pure, and customer oriented due to proper use of available facilities, suitable utilization of new sources for producing goods, and introduction of desirable services with suitable quality. Using philosophies like Lean manufacturing and employing tools such as technology, firms can establish an efficient and stable system to improve their weak points and protect their strong points (recoverable areas), enabling them to continuously identify their planning priorities and recover their recoverable areas by using corrective actions, resulting in gradually passing organizational transcendence levels and improving their efficiency.

Among various tools for performing Lean manufacturing assessments, technology has remarkably allowed the obtaining of world class function as well as recovering job function. In addition, technologies have stimulated immense attention as they provide a powerful tool for continuous recovering which is the focus of many organizations and firms. The general aim of technology is the reorganization of tools, techniques, crafts, systems, and methods of organization together with the method of applying them to solve a problem or create an artistic perspective. In order to reach this goal, interactions should be used to identify the cause and effect of relationships between technology and Lean manufacturing. In this way, the main problem of an organization originates from little or weak identification of the cause and effect structure between technology and Lean manufacturing, while a systemic approach can solve this issue. The relationship observation from systemic sight is in consistence with primary assumptions of its development. Systemic approach suggests that all different aspects and organizational areas are related to each other, and one cannot recover an area without affecting other areas even in a whole-area recovery. On the other hand, among several observable variables and their relations, special cause and effect loops are prevailing in determining the general behavior of a system.

Challenges from global competitors during the past two decades have prompted many manufacturing firms to adopt new manufacturing approaches [2–4]. In particular prominent among these is lean manufacturing [5, 6]. With the notable exception of [7], there is relatively few published evidence about the implementation of lean practices and its effective factors. A majority of articles on the topic of lean manufacturing systems focus on the relationship between implementation of lean and technology.

However, conceptual research continues to stress the importance of the effect of technology on lean manufacturing programs. We specifically examine the relationship between the factor of technology and lean systems. This contextual factor has been suggested as a possible facility to implement lean manufacturing systems.

THEORETICAL BASES

Lean Manufacturing

Lean manufacturing or Lean production, often simply, "Lean," is a production practice that considers the expenditure of resources for any goal rather than the creation of value for the end customer, which is considered wasteful and a target for change. Working from the customer's perspective who consumes a product or service, "value" is defined as any action or process that he would be willing to pay for. Basically, lean is centered on preserving the value, with less work. Lean manufacturing is a generic process management philosophy derived mostly from the Toyota Production System (TPS) (hence, the term Toyotism is also prevalent) and identified as "Lean" only in the 1990s [6, 8, 9]. The original Toyota seven mood reduction is a well-known approach to improving overall customer value, but there are varying perspectives on how this is best achieved. The steady growth of Toyota, from a small company to the world's largest automaker, has focused attention on how this was achieved [10]. Vinodh and Balaji [11] reported a study which is carried out to assess the leanness level of a manufacturing organization in which a leanness measurement model has been designed, the leanness index has been computed, and a computerized decision support system has been developed. The model computes the fuzzy leanness index, Euclidean distance and identifies the weaker areas which need improvement.

Lean manufacturing is a variation in the theme of efficiency based on optimizing flow; it is a present-day instance of the recurring theme in human history toward increasing efficiency, decreasing waste, and using empirical methods to decide what matters, rather than uncritically accepting preexisting ideas. As such, it is a chapter in the larger narrative, that also includes such ideas as the folk wisdom of thrift, time and motion study, Taylorism, the Efficiency Movement, and Fordism. Lean manufacturing is often seen as a more refined version of earlier efficiency efforts, building upon the work of earlier leaders such as Taylor and Ford and learning from their mistakes [12]. Leanness assessment using multigrade fuzzy approach was proposed by Vinodh and

Suresh [13]. In the research, a leanness measurement model incorporated with multigrade fuzzy approach was designed. This is followed by the substitution of the data gathered from a manufacturing organization. After the computation of leanness index, the areas for leanness improvement have been identified. They indicated that the approach contributed in the project could be used as a test kit for periodically evaluating an organization's leanness. The performance of a lean cell that implements lean goals under uncertainty was investigated by Deif [14]. The investigation is based on a system dynamics approach to model a dynamic lean cell. Backlog is used as a performance metric that reflects the cell's responsiveness. The cell performance is compared under certain and uncertain external (demand) and internal (machine availability) conditions. He explores the effect of the delay associated with the proposed capacity policies and how they affect the lean cell performance. A model for measuring adherence to lean practices for automotive part suppliers and to assess the relationship between the firm performance and the adoption of lean principles was proposed by Sezen [15]. Their model applied to a large number of automotive part suppliers in Turkey and data was collected from 207 automotive part suppliers by using the computer-aided telephone interview method. Validity and reliability tests of the developed model of leanness are realized through exploratory and confirmatory factor analyses. The study shows that, in general, Turkish automotive part suppliers are performing their internal production in compliance with the lean manufacturing principles. Furthermore, they found that there is a significant relationship between adaptation of lean principles by the supplier firms and their performance.

Technology

Technology is the usage and knowledge of tools, techniques, crafts, systems, or methods of organization in order to solve a problem or create an artistic perspective. Technologies significantly affect the human's (as well as other animal species') ability to control and adapt to their natural environments. Human use of technology began with the conversion of natural resources into simple tools. The prehistoric discovery of the ability to control fire increased the available sources of food and the invention of the wheel helped humans in travelling and the control of their environment. Recent technological developments, including the printing press, the telephone, and the Internet, have reduced physical barriers to communication and allowed humans to interact freely on a global scale. However, not all technology has been used for peaceful purposes; the development of weapons of ever-increasing destructive power has progressed throughout history, from clubs to nuclear weapons.

Technology has affected society and its surroundings in a number of ways. In many societies, technology has helped the development of more advanced economies (including today's global economy) and has allowed the rise of a leisure class. Many technological processes produce unwanted by-products, known as pollution, and deplete natural resources to the detriment of the Earth and its environment. Various implementations of technology influence the values of a society, and new technology often raises new ethical questions. Examples include the rise of the notion of efficiency in terms of human productivity, a term originally applied only to machines, and the challenge of traditional norms [16].

System Dynamics

System dynamics is an approach to understanding the behavior of complex systems over time. It deals with internal feedback loops and time delays that affect the behavior of the entire system. What makes the use of system dynamics differ from other study methods of complex systems is the use of cause and effect diagrams and feedback loops in addition to the stock and flow diagram. These elements help one to argue how seemingly simple systems display baffling nonlinearity.

The field of system dynamics was developed in the early 1960s, initially from the work of Jay Forrester at MIT. Causal loops capture mental models and relationships in a system.

Dynamic systems modeling in educational system was proposed by Groff [17]. He mentioned that applying this tool to educational policy analysis offers insights into the hidden dynamics of the current system and can be an invaluable tool in designing future scenarios. He explored underlying dynamics of the current US educational system using system dynamics modeling and offered an analysis of this tool and its practical application in the US educational system through a case study on the US state of Rhode Island in the 2007-2008 school year.

The scenario technique is a strategic planning method that aims to describe and analyze potential developments of a considered system in the future. Its application consists of several steps, from an initial problem analysis over an influence analysis to projections of key factors and a definition of the scenarios to a final interpretation of the results. The technique itself combines qualitative and quantitative methods and is an enhancement of the standard scenario technique [18]. They used the numerical values gathered during the influence analysis and embedded them in a system dynamics framework which yields a mathematically rigorous way to achieve predictions of the system's future behavior from an initial impulse and the feedback structure of the factors.

System dynamic model approach for urban watershed sustainability study was studied by Feng [19]. He investigated the dynamic interactions between natural environment and human society to model long-term trends in environmental impact and sustainable development. The data include 21 environmental, social, and economic indicators for five counties. The data show that, within the study area, population has increased by an annual average of 6.4% with a range from −7.9% to 20.7% over 30 years. To project the future of environmental sustainability, a system dynamic model was established. Results suggest that population will remain stable, in 2010.

Stocks and flows describe how a system is connected to feedback loops, which create the nonlinearity that can be found so frequently in modern day problems. Computer software is used to simulate a system dynamics' model for the considered situation. In such a model, running "what if" simulations for testing certain policies can greatly aid in understanding how a system changes over time [20, 21].

Cause and Effect Diagrams

Causal loop diagrams are used to capture mental models and represent interdependencies and feedback processes in a system. All dynamics arise from the interaction of only two types of feedback loops, positive and negative. Positive loops tend to reinforce or amplify the occurring events in the system, while negative loops counteract and oppose change.

Stock and Flow Diagrams

Stock and flow diagrams are a central part of the dynamic system theory. They are used to capture the stock and flow structure of systems. A stock is defined as a supply accumulated for future use, while a flow describes how the stock increases and decreases by inflows and outflows. The dynamics of the system are brought forth by examining the differences between inflows and outflows to a stock.

THE STEPS IN THE MODELING OF A DYNAMIC SYSTEM

Logical steps in the modeling of a dynamic system are as follows.

(i)   Definition of problems that need to be solved and the results that need to be achieved.

(ii)   Analysis of the problem with the help of cause-and-effect diagrams.

(iii   )Formulation of the model structure.

(iv)   Collection of information, initial values, and the basic data needed for the construction of the model from existing data and/or discussion with conductors or designers who have the knowledge and experience of the system under study. The initial values, the state values, the constant values, and the data related to the policies can also be considered among these.

(v)   Investigation of model validity under certain conditions to ensure model validity.

(vi)   Employment of the model in testing various policies to reach the most suitable results [20].

BENEFITS OF ANALYZING TECHNOLOGY EFFECTIVENESS OF LEAN MANUFACTURING, USING DYNAMIC SYSTEMS

There are three significant benefits for using and developing system dynamics in the relationship between technology and lean manufacturing, which may be considered as below.

Conversion of Unidirectional to Bidirectional Causality

In using technology, most organizations consider unidirectional causal relations. The use of causal loops alone is seen as problematic and in contradiction with reality. Instead of a causal relationship, this model believes that the relationship is more of an interdependence or bidirectional causality, relying on the fact that causal relationships are seldom unidirectional in the real world.

In the proposed dynamic model the effect of new technology is considered on lean manufacturing. These effects act as bidirectional, meaning that lean manufacturing also affects technology.

Considering Time Dimension in Cause and Effect Relationships

A common problem arises from the fact that time dimension is not considered as a part of Lean manufacturing, while in some cause and effect relationships a time lag does exist between the cause and effect. This time lag is not shown by technology effectiveness (TE) of lean manufacturing since it measures the cause and effect at the same time. Simply looking at different measures simultaneously is not enough, and the linkages between them must also be understood.

In the provided dynamic model in this paper, as it is in reality, a time lag exists between new technology and lean manufacturing.

A Mechanisms for Validation

The analysis of technology effectiveness of lean manufacturing using system dynamics provides the mechanism for maintaining the relevance of defined measures. The problem for managers is usually reducing the list of possible measures to a manageable (and relevant) set rather than identifying what could be measured. Thus, the advantage of checking a few numbers may become a disadvantage if the right numbers are not selected for lean manufacturing.

Furthermore, the analysis of company strategy based on the lean manufacturing approach considers the causal relationships between performance variables only in qualitative terms. This implies that managers should rely on mental simulations and heuristics in order to quantify the results of their strategy and, hence, evaluate its efficiency and effectiveness. This task is even tougher when the company system is characterized by a high degree of complexity, nonlinear relationships among variables, and delays between causes and effects.

The validity of technology effect on causal relationships between the variables has also been questioned by system dynamics. In particular, it has been demonstrated that the hypothesized links between implementing new technology and profit may be not confirmed in reality. For instance, it has been remarked that the commonly assumed causal relationship of productivity and profit may not have any empirical evidence. On the contrary, it may be seen that the costs of policies aimed to increase productivity are higher than the related benefits, both in short and long term. For such reasons, not considering the effects of technology may lead management to mistakes.

MODELING PROCESS

The purpose of dynamic system modeling is to establish the relation between the various variables which build the system and are used to analyze decision making policies in the realm under study. The cause-and-effect diagram is an essential tool which helps in modeling the real world in the form of feedback links. The effectiveness variables in this relation are as follows.

Cause-and-Effect Diagram Modeling

The cause-and-effect diagram of this model shows the relation between technology and lean manufacturing. The key effectiveness variables in this relation are new technology, employee productivity, innovation in processes, customer satisfaction, financial sources, complexity of industry, and lean manufacturing. Expanded cause-and-effect relations between technology and lean manufacturing can be described as follows.

(i)   The relationship between new technology implementation and innovation in processes.

(ii)   The relationship between new technology implementation and interaction between employees.

(iii)   The relationship between new technology implementation and the cost to change technology.

(iv)   The relationship between new technology implementation and lean manufacturing.

In this paper due to the expanded relations defined above, the relationships between the model's variables are used in drawing the model. More description is provided as follows.

It is a common belief among enterprises that reaching lean results, regarding customer satisfaction, employee productivity, and innovation in processes, requires effective technology, which itself begins with putting together new program needs and is expanded with the compilation of organizational changes. Therefore, organizational changes and new program needs are effective on all model variables. An organization can attain productivity regarding employees only when it elects suitable approaches in improving the interaction between employees; for this reason, technology is related to employee productivity. By achieving lean results in the field of customer satisfaction, the utilization of suitable approaches is possible in innovational processes such as design and expansion of products, delivery of products and services, and management of communication with customers. Therefore customer satisfaction is related to the process. Reaching lean results in the field of financial sources requires reaching improvement in customer satisfaction and employee productivity. Therefore, the criteria of key operation results are in relation with customer satisfaction and employee productivity.

A simplified and stylized version of the qualitative model that is the end result of the first modeling phase is shown in Figure 1. In this causal loop diagram, nine interconnected feedback loops are shown and together determine the dynamic behavior of the model. These are labeled R1···R9 with the "R" standing for "reinforcing" or positive feedback loop and B1, B2 with the "B" standing for "balancing" or negative feedback loop.

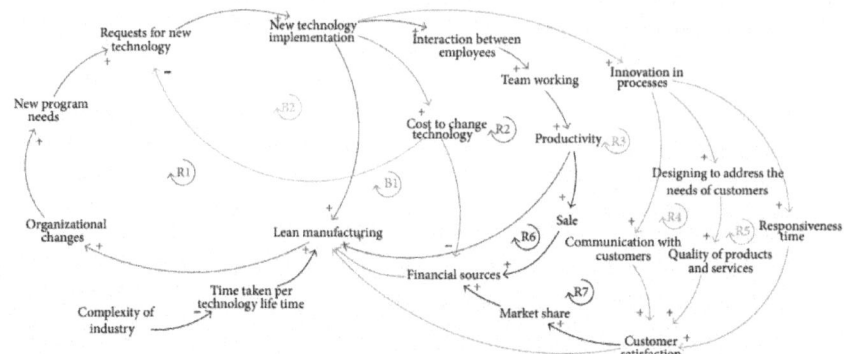

Figure 1: Cause-effect diagram based on the relationship between technology and lean manufacturing.

Some of the relations between variables are reviewed in the past articles [22, 23] and the rest is based on excellence models like the EFQM model [24].

R1: The New Technology Implementation Loop. A pertinent observation in the modeling phase was that new program needs increase as a result of an increase in organizational changes. Thereafter implementing new technology is made essential, leading to an increase in the request for new technology. Finally, implementing new technology leads to better lean manufacturing. As a result, technology is a tool that can directly affect lean manufacturing.

R2: The Productivity Loop. Interaction between employees increases as a result of implementing new technology. Thereafter employee productivity increases, leading to an increase in team working. Finally, employee productivity leads to better lean manufacturing. As a result, technology is a tool that affects productivity, and therefore productivity can affect lean manufacturing.

R3: The Communication with Customers Loop. Improvement and innovation in processes increase as a result of implementing new technology. Thereafter customer satisfaction increases, leading to better communication with customers. Finally, customer satisfaction leads to better lean manufacturing. As a result, technology is a tool that affects customer satisfaction, and therefore customer satisfaction can affect lean manufacturing.

R4: The Quality of Products and Services Loop. Improvement and innovation in processes increase as a result of implementing new technology. Thereafter customer satisfaction increases, leading to better quality of products and services. Finally, customer satisfaction leads to better lean manufacturing. As a result, technology is a tool that affects customer satisfaction, and therefore customer satisfaction can affect lean manufacturing.

R5: The Response Time Loop. Improvement and innovation in processes increase as a result of implementing new technology. Thereafter customer satisfaction increases, leading to better response time. Finally, customer satisfaction leads to better lean manufacturing. As a result, technology is a tool that affects customer satisfaction, and therefore customer satisfaction can affect lean manufacturing.

R6: The Productivity-Financial Sources Loop. Interaction between employees increases as a result of implementing new technology. Thereafter employee productivity increases, leading to an increase in team working. Thereafter, the number of sales increases as a result of an increase in productivity, which leads to a rise in financial sources. Finally, financial sources lead to better lean manufacturing. As a result, technology is a tool that affects productivity, and therefore productivity can affect financial sources.

R7: The Customer Satisfaction-Financial Sources Loop. Improvement and innovation in processes increase as a result of implementing new technology. Thereafter customer satisfaction increases, leading to better response time, quality of products and services and communication with customers. Thereafter, marketing shares increase as a result of an increase in customer satisfaction, which leads to a rise in financial sources. Finally, financial sources lead to better lean manufacturing. As a result, technology is a tool that affects customer satisfaction, and therefore customer satisfaction can affect financial sources.

B1: The Cost of Changing the Technology-Financial Source Loop. The cost to change technology increases as a result of new technology implementation. Thereafter financial sources decrease, leading to an increase in the cost of changing technology. Finally, a decrease in financial sources leads to a decrease in lean manufacturing. As a result, the cost to change technology is a tool that affects financial sources.

B2: The Cost to Change Technology-Requests for the New Technology Loop. Requests for new technology decrease as a result of an increase in the cost to change technology. Thereafter implementation of new technology decreases, leading to a decrease in the cost to change technology. As a result, cost is a criterion that affects requests for new technology.

Moreover, a short time pertinent observation in the modeling phase was that expenditures increase as a result of an increase in new technology, but in the long run new technology led to an increase in productivity and innovation in processes. Thereafter sale and customer satisfaction increased. Finally, sale and customer satisfaction led to better financial sources. In the short run, costs would increase, but in the long run, costs will decrease, leading to an increases in benefit. Therefore, as a result, it is necessary to consider technology in organizations' short and long time effects. In this diagram two parallel lines (II) are employed for showing delay in the relationships between variables.

Stock and Flow Diagram Modeling

In order to describe the relations between the variables and to investigate various scenarios, interviews with experts were carried out.

The stock and flow diagram based on the relationship between technology and lean manufacturing is presented in Figure 2.

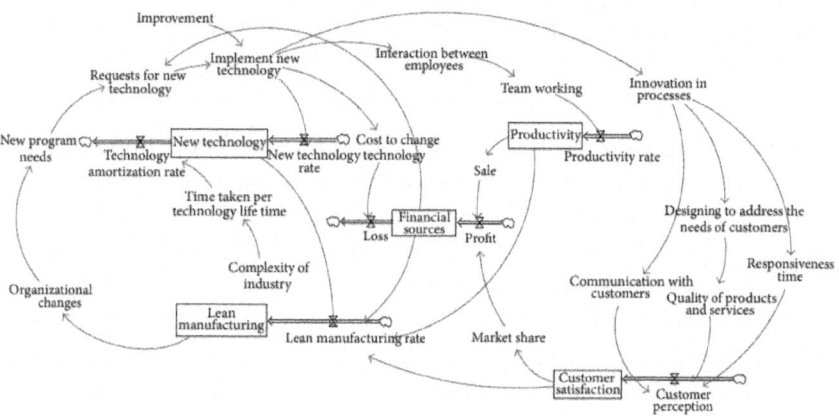

Figure 2: Stock and flow diagram based on the relationship between technology and lean manufacturing.

The developed model is performed by using the Vensim PLE Software. Time unit is set to a year and the model is run for 13 years, starting from 2008.

To study the trend of organizational development, one can define levels during the time of new technologies' effects. In this paper, we have defined the levels as new technology, productivity, customer satisfaction, financial sources, and lean manufacturing. These levels indicate organizational changes due to the complexity of industry, during the run time.

The results of the simulation of "lean manufacturing," "financial sources," "customer satisfaction," and "productivity" levels are shown in Figure 3.

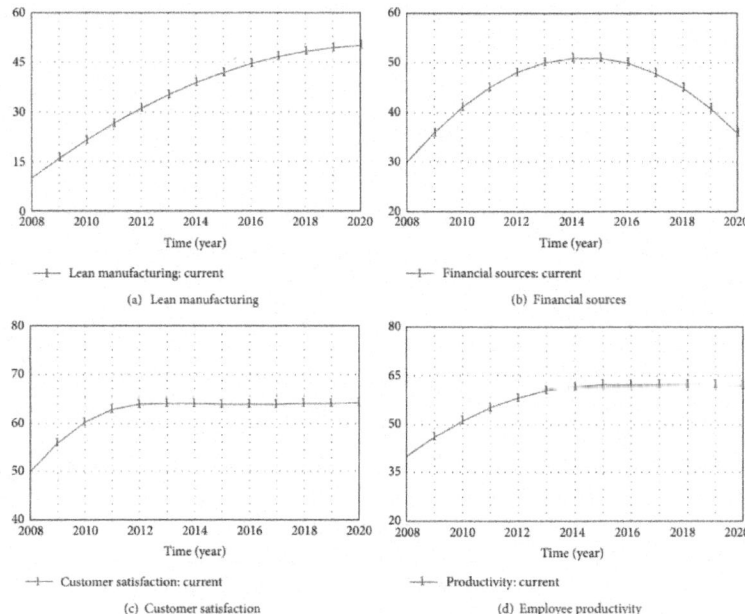

Figure 3: Behaviors of prominent variables obtained through simulation.

It can be seen that the behavior of Figures 3(a), 3(c), and 3(d) is goal seeking. This may be explained by the increase in productivity and customer satisfaction due to the new technology implementation. Consequently, lean manufacturing also improves during this time.

Furthermore, it can be seen that the behavior of Figure 3(b) in initial stages increases but shows decreasing behavior later on. In the short time, this could be explained by the rise of expenditures due to the new technology implementation, but in the long run the implemented new technology causes an increase in productivity and innovation in processes, which consequently causes a decrease in expenditures.

For the linking process between these variables, the kind of auxiliary variables in the Vensim PLE Software is set to "lookup," and is used for the relationship between the variables of the model. For example, to link between the "new program needs" variable and the criteria "organizational changes" and "new program needs," the type of the variable is set to "lookup," which shows that the trend of cause changes based on effect changes. This is expressed in Figure 4. In this diagram, "organizational changes" is placed on the X-axis and "new program needs" is placed on the Y-axis.

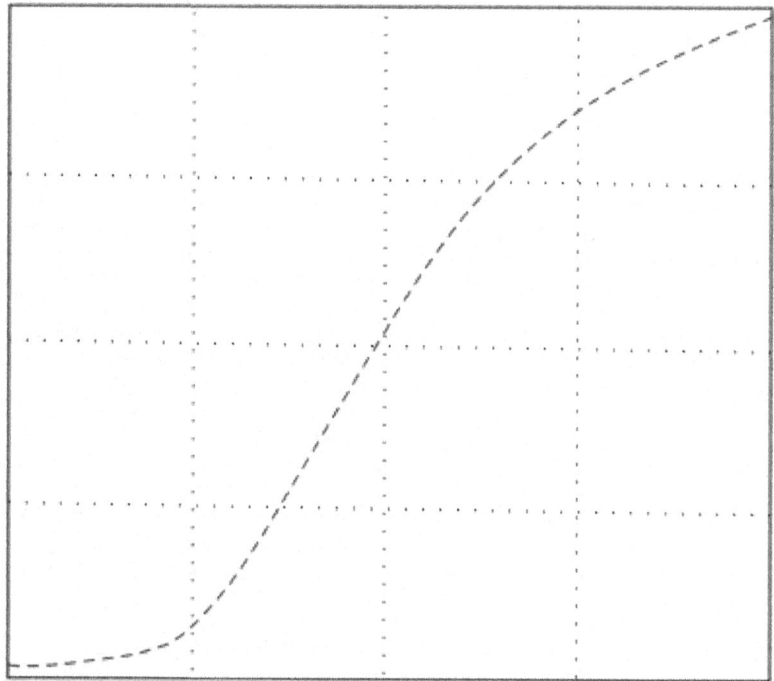

Figure 4: "New program needs" variable.

Performance Tests of the Proposed Model

In order to validate dynamic models, conventional tests such as boundary efficiency, unit's consistency, parameter evaluation, structure evaluation, cumulative error, and the extreme value test are carried out.

(i)   The boundary efficiency test suffices parameters and causal loops in the model according to the purpose. This problem is verified in the modeling phase after reaching results in the interaction with experts and has more emphasis on model logic.

(ii)   The unit's consistency test emphasizes on the equality of units in the model and is verified by the software.

(iii) The parameter evaluation test emphasizes on a correct definition of variable's initial amounts and on the base of objective data or anticipation. For this purpose, all used data in the simulation is taken from existing chronological data or has been adjusted on the base of anticipation.

(iv)   The structure evaluation test considers the compatibility of the model behavior with its structure. This problem necessitates that variable behavior in negative and positive feedback, in the simulated model, must be orderly exponential and seeking its goal. On this base, as seen in the causal diagram, the variables from negative feedback loops and their behavior in the simulated model must be goal seeking, which can be clearly seen in Figure 5.(v)The cumulative error test verifies that the results of the simulation are not sensitive to the time unit. For example, if the time unit was set to one year, changing it to six months should not alter the results.

(vi)   The extreme value test emphasizes on the model's resistance in limited conditions, meaning that the model must show its expected behavior under any circumstances, even under the change of policies or entry amounts. For example, even if the initial amounts of the external variables are increased or decreased greatly, the internal variables' changes must be in their genuine range. This matter is also considered in the proposed model and is verified regarding limited conditions. For example, the amount of the variable "lean manufacturing" is tested while the "complexity of industry" variable is changed from 0 to 100. The result is shown in Figure 6.

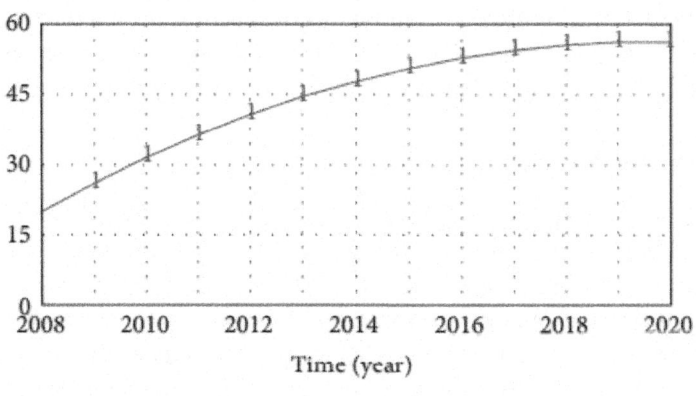

—+— Market share: current

Figure 5: Behaviors of "marketing share" variable.

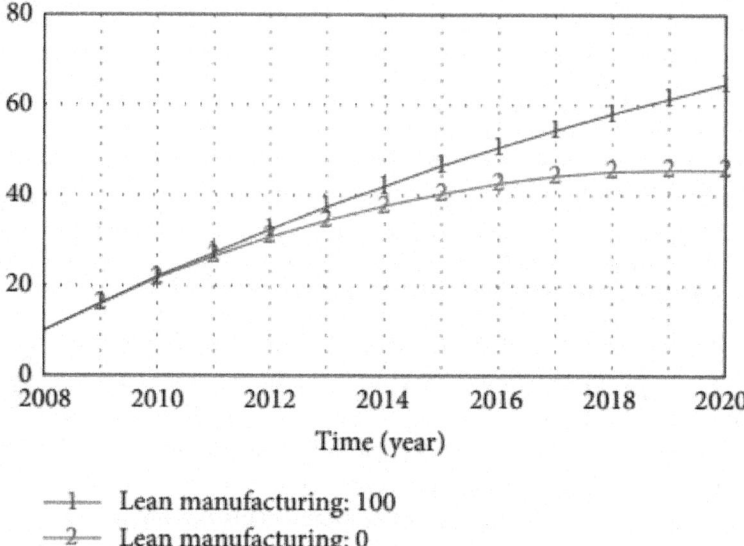

—1— Lean manufacturing: 100
—2— Lean manufacturing: 0

Figure 6: Behaviors of "lean manufacturing" variable.

It could be seen that behavior of the "lean manufacturing" variable has not changed in its limited conditions. Consequently, by model verification, it is possible to make scenarios to analyze the results and subsequently make decisions that are discussed as follows.

POLICY MAKING

In the proposed dynamic model an "improvement" variable is defined for the "implementing new technology" variable in order to evaluate different policies (the improvement variable is determined based on the future goals and policies of the organization). We will now discuss and compare three different policies and scenarios to find the decisions by which the firm may achieve its objectives. "Lean manufacturing" and "financial sources" are the main variables which we must pay special attention to. We will consider the following three scenarios.

Policy 1. We name the first scenario the "low improvement of technology approach." Therefore, we assume a low value of improvement in the variable, considering its logical value to be 10%.

Policy 2. We name the second scenario the "middle improvement of technology approach." Therefore, we assume an average value of improvement in the variable, considering its logical value to be 25%.

Policy 3. We name the third scenario the "high improvement of technology approach." Therefore, we assume a high value of improvement in the variable, considering its logical value to be 50%.

The results of applying each of the above policies to the "lean manufacturing" and "financial sources" variables are shown in Figures 7 and 8.

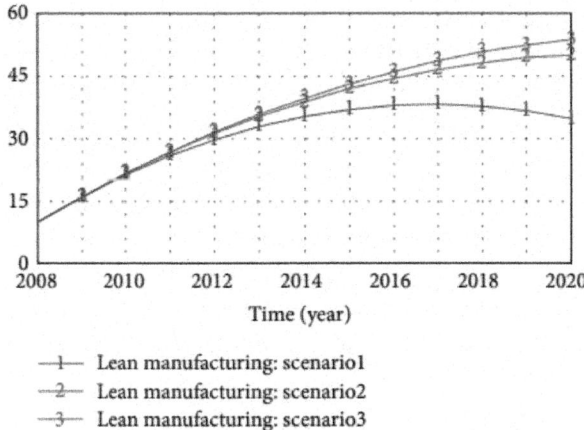

Figure 7: The results of applying the three policies for the "lean manufacturing" variable.

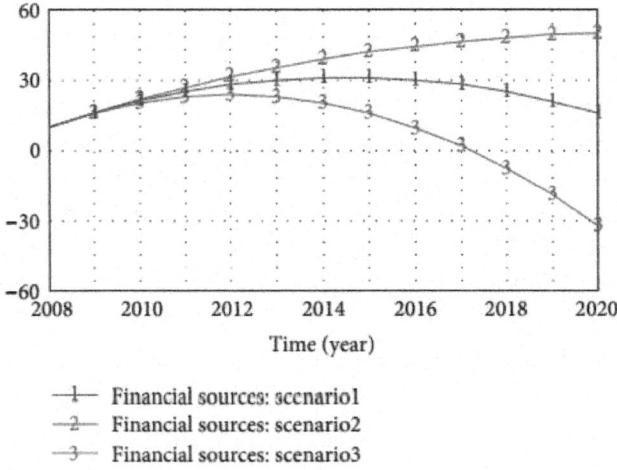

Figure 8: The results of applying the three policies for the "financial sources" variable.

As seen in Figures 7 and 8 the "lean manufacturing" and "financial sources" variables have a goal seeking behavior in all three scenarios.

In Figure 7, results of the simulation show that "lean manufacturing," in all three scenarios, increases faster in the first years compared to later years and finally decreases in the end. In comparison between the three scenarios, it is observed that the third scenario has a better following of the mentioned trend than the other two scenarios.

In Figure 8, the results of the simulation show that "financial sources" in all three scenario increase at first but decrease in later years. In comparison between the three scenarios it is observed that the second scenario has a better following of the mentioned trend than the other two scenarios.

Since the trend of the second scenario in the "lean manufacturing" variable is nearly at the same level of the trend the third scenario and in the "financial sources" variable it has a better following of trend compared to the other two scenarios, hence it is considered as the best choice among the three scenarios, meaning that new technology implementation must be proportionate to its cost and income. In other words, if the improvement in lean manufacturing is lower in proportion to its increase in cost in a scenario, that scenario is declined, as is the third scenario in this paper.

Therefore, the second scenario should be chosen in enterprises, as it obtains more benefits and allows technology to be useful and valuable in lean manufacturing.

CONCLUDING REMARKS

This research provides a model with systemic function in order to analyze the relationship between technology and lean manufacturing in enterprises. The developed model is highly suitable for describing and analyzing organizations. We have introduced a new integrated dynamic approach which investigates technology effectiveness of lean manufacturing in enterprises and selects the best policy among the enforceable policies. Consequently, we have indicated the effects of powerful factors on organizational results by using the proposed dynamic model and have obtained and analyzed the trend of changes in terms of different values by utilizing the Vensim PLE Software.

Analysis of technology effectiveness of lean manufacturing, considering the dynamic behavior of the system, provides a group of advantages. The most important are as follows.

(i)   Simulating the effect of important factors on enterprise results.

(ii)   Performing the "what if" analysis to learn from future potential threats and scenes.

(iii) Capability of visual representation for relations between the values of the model.

(iv)   Reducing the risk of performing future plans through simulating and studying the results and the conclusions of different policies.

Also, obtaining the dynamic model provides advantages for recovering evaluations, including time dimension between cause and effect. With respect to the evaluations made based on the developed model, it is indicated that the effects of change in powerful values on the results occur simultaneously. However, the effects appear during the next period rather than the same period. Therefore it is fair to say that the proposed dynamic model is a suitable tool for modeling the situation of organizations and predicting the effect of their existing strategies.

The following fields may provide a base for future research opportunities.

(i)   More complete performance tests of the developed model.

(ii)   Performing more simulations for different policies with different analysis of their results which would lead to a more exact conclusion.

(iii) In the developed model, we have provided the relationship between technology, employees, processes, and lean manufacturing. A future research may be to include more relations by introducing more variables, for example, the partnerships, outsources, and knowledge management variables.

(iv)   Inviting experts of manufacturing systems and system dynamics for further research on the developed model may lead to the improvement of the relationships and equations of the model.

(v)   Modeling and measuring alone are not enough for organizations. These techniques must be considered in social-economical systems.

The developed model may be combined with organization policies to raise its efficiency. Also, discussing and developing the model by implementing it at different positions may provide guidance for future investigations.

CONFLICT OF INTERESTS

The authors state that they are in no way associated with or financially related to any mentioned commercial identities in this paper. Any mention of commercial identities is for illustrative purposes only and should not interfere with matters such as conflict of interests.

REFERENCES

1.   Aberdeen Group, "The role of technology in lean manufacturing," August 2007.

2. R. W. Hall, Attaining Manufacturing Excellence: Just-in-Time, Total Quality, Total People Involvement, Dow Jones-Irwin, Homewood, Ill, USA, 1987.

3. J. R. Meredith and R. Mctavish, "Organized manufacturing for superior market performance," Long Range Planning, vol. 25, no. 6, pp. 63–71, 1992. ·

4. S. Goyal and S. Grover, "A comprehensive bibliography on effectiveness measurement of manufacturing systems," International Journal of Industrial Engineering Computations, vol. 3, no. 4, pp. 587–606, 2012. ·

5. J. P. Womack and D. T. Jones, Lean Thinking: Banish Waste and Create Wealth in Your Corporation, Simon & Schuster, New York, NY, USA, 1996.

6. J. P. Womack, D. T. Jones, and D. Roos, The Machine That Changed the World, Harper Perennial, New York, NY, USA, 1990.

7. R. E. White, J. N. Pearson, and J. R. Wilson, "JIT manufacturing: a survey of implementations in small and large U.S. manufacturers," Management Science, vol. 45, no. 1, pp. 1–15, 1999. ·

8. Mahfouz, J. Shea, and A. Arisha, "Simulation based optimization model for the lean assessment in SME: a case study," in Proceedings of the Winter Simulation Conference (WSC '11), pp. 2403–2413, Phoenix, Ariz, USA, December 2011. · ·

9. M. Taleghani, "Key factors for implementing the lean manufacturing system," Journal of American Science, vol. 6, no. 7, pp. 287–291, 2010.

10. M. Holweg, "The genealogy of lean production," Journal of Operations Management, vol. 25, no. 2, pp. 420–437, 2007. · ·

11. S. Vinodh and S. R. Balaji, "Fuzzy logic based leanness assessment and its decision support system,"International Journal of Production Research, vol. 49, no. 13, pp. 4027–4041, 2011. · ·

12. D. Bailey, "Automotive news calls Toyota world no 1 car maker," 2008.

13. S. Vinodh and K. C. Suresh, "Leanness assessment using multi-grade fuzzy approach," International Journal of Production Research, vol. 49, no. 2, pp. 431–445, 2011. · ·

14. M. Deif, "Dynamic analysis of a lean cell under uncertainty," International Journal of Production Research, vol. 50, no. 4, pp. 1127–1139, 2012. · ·

15. Sezen, I. Karakadilar, and G. Buyukozkan, "Proposition of a model for measuring adherence to lean practices: applied to Turkish automotive part suppliers," International Journal of Production Research, vol. 50, no. 14, pp. 3878–3894, 2012. ·

16. "Definition of technology," Merriam-Webster, 2007.

17. J. S. Groff, "Dynamic systems modeling in educational system design & policy," New Approaches in Educational Research, vol. 2, no. 2, pp. 72–81, 2013.

18. Brose, A. Fügenschuh, P. Gausemeier, I. Vierhaus, and G. Seliger, "A system dynamic enhancement for the scenario technique," ZIB-Report 13-24, 2013.

19. H. Feng, D. Yu, Y. Deng, M. P. Weinstein, and G. Martin, "System dynamic model approach for urban watershed sustainability study," International Journal of Sustainable Development, vol. 5, no. 6, pp. 69–80, 2012.

20. J. D. Sterman, Business Dynamics: Systems Thinking and Modeling for a Complex World, McGraw-Hill, Irwin, Pa, USA, 2000.

21. T. Daniel and L. Bengt, "Aggregate analysis of manufacturing systems using system dynamics and ANP," Computers and Industrial Engineering, vol. 49, no. 1, pp. 98–117, 2005. · ·

22. S. K. Somavarapu, System dynamics approach to understand the role of information technology in the evolution of next generation integrated product development systems, Massachusetts Institute of Technology, 2005.

23. H. Shafiul, M. Al-Hussein, and P. Gillis, "Advanced simulation of tower crane operation utilizing system dynamics modeling and lean principles," in Proceedings of the Winter Simulation Conference (WSC '10), 2010.

24. S. M. Dehghani, M. S. Owlia, B. Kiani, and K. Noughandarian, "Analysis of EFQM excellence model using system dynamics," Iranian Journal of Management Sciences, vol. 3, no. 12, pp. 65–82, 2009.

Chapter 8

MULTI-AGENT BASED DISTRIBUTED MANUFACTURING

J. Li[3], J.Y H. Fuh[1], Y.F. Zhang[2], and A.Y.C. Nee[3]

[1]University of California, Los Angeles (UCLA), Los Angeles, California, United States.
[2]National University of Singapore, Lower Kent Ridge Rd, Singapore
[3]University Campus STeP Ri Slavka Krautzeka 83/A 51000 Rijeka, Croatia

INTRODUCTION

Agent theory is developed from distributed artificial intelligence, which is regarded as a prospective methodology suitable for solving distributed complex problems, and it has been applied in many areas including manufacturing engineering. In this chapter, some basic issues for agent theory are described and an example of one agent-based distributed manufacturing system is presented. 1.1 Agent and multi-agent system Jennings and Wooldridge (Jennings and Wooldridge 1998) have defined an agent as "a computer system situated in some environment and capable of autonomous action in this environment, in order to meet its design objectives". Some of the main properties of agents are autonomy, socialability, reactivity, and proactiveness (Wooldridge and Jennings 1995):

- Autonomy: Autonomy characterizes the ability of an agent to act on its own behalf. Agents can operate without direct intervention of humans or other agents, and have a some kind of control over their actions and internal states (Castelfranchi 1995).

- Sociability: Agents can interact with other agents via agent communication languages (Gensereth and Ketchpel 1994)

- Reactivity: Agents can perceive the changes of their environment, which may be the physical world, a collection of other agents, the Inter net, and other fields, and respond to make the related decision accordingly in real time.

- Proactiveness: Agents do not only act in response to their environment, but also exhibit goal-directed behavior by taking the initiative

All of these properties are necessary for agents to act as autonomous, loosely coupled and self coordinating entities in an open distributed system; which forms a multi-agent based system (MAS). A MAS consists of a group of agents that play individual roles in an organizational structure (Weiss 1999). The most important characteristic of MAS is the agents' capabilities of communication and cooperation, which make them to interact with other agents to achieve their individual objectives, as well as the common goals of the system (Wooldridge and Jennings 1995). Other important characteristics of the agent- based systems include scalability, modularity and re-configurability. In an MAS model, every agent is a representative of a functional cell. For in- stance, in order to agentify a complex system, it will be divided into some sub- systems, each of which is further encapsulated into an agent. Each agent con- quers its individual problem, and cooperates with other related agents to solve the whole problem. In the distributed system modeling, an agent is the repre- sentative of a distributed cell which solves its own problems and can cooperate with other agents to fulfill a task if necessary. A comprehensive book on multi- agent theory can be found in (Weiss 1999).

The architecture of MAS

The architectures of multi-agent based systems provide the frameworks within which agents are designed and constructed (Shen 2002). Similar with the or- ganization of the distributed manufacturing system, there are three types of architecture for multi-agent based systems, which are hierarchical (A), heterar- chical (B) and hybrid structures (C), as shown in the figure 1. Ner.

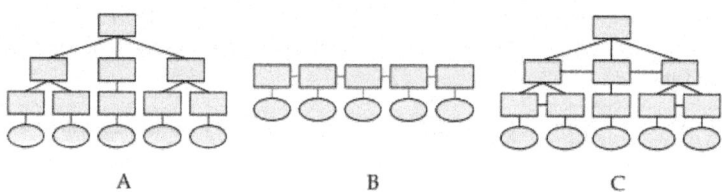

A   B   C

Figure 1: Architecture of MAS.

In the hierarchical architecture (A), the agents are connected with layered rela-tionship and all of the control modules are organized into a hierarchical man- Each agent will have only one direct supervisor at its directly upper layer and several subordinate agents at its directly lower layer. The agent executes the commands and plans only from its supervisor agent and gathers the feed-back information from its subordinate agents. The main advantage for this structure is that global optimization can be achieved possibly as the complete information and status of the system can be collected by the agent at the highest layer; while the main disadvantages resides in less adaptability and reliability because the system may be malfunction once the central controller agent breaks down.   Heterarchical architecture (B) is another different style compared with the pre- vious one because there is no central controller in this kind of structure and the   relationship of the agents is peer to peer. Each of the agents is autonomous and   has its own decision-making mechanism. The cooperation work among agents   is to be realized by negotiation: the related agents will negotiate and make   tradeoff for a variety of factors. The advantage for this type architecture is its   high robustness because breakdown of one agent will not influence others and   the rest can still work. The main problem for this architecture lies in the diffi- culty to achieve a global optimization as no single agent can collect the full in- formation and status from others. Furthermore, another shortcoming is that   the execution efficiency is relative low in such framework because the negotia- tion process may be inefficient and less effectiveness, especially for those tasks   need to be completed by several cooperative agents.   The third type (C) is the hybrid architecture, which can be regarded as a com- promise of the above two kinds. The hierarchy of the system enhances its effi- ciency and effectiveness on a global basis while achieving some advantages of   the heterarchical architecture to keep the good adaptability and autonomy. In   this architecture, the agents at the lower level are also intelligent and have   some degree of autonomy, which can be viewed as a heterarchical structure.   But the agents also have their upper layered supervisor agent, which can col- lect the information and distribute tasks to some capable subordinate agents.   As the upper level supervisor agent can get a global view for its subordinate   agents, some global optimal decision can be achieved. At the same time, as the   lower level agents are autonomous, some decisions can be made locally and   will not impact other agents, which can improve the robustness and adaptabil- ity of the whole system.

The coordination methodology for MAS

The methodology of negotiation and coordination is one of the bases for ef-fective management and control in a distributed system. Presently, the well-known Contract Net Protocol (CNP) (FIPA 1997; FIPA 2000(1)) is adopted as the coordination and negotiation protocol in most of multi-agent systems. CNP method was proposed by smith (Smith 1980; Davis and Smith 1983;   Smith 1988) and recommended by FIPA(The Foundation for Intelligent   Physical

Agents)(FIPA 2000(1); FIPA 2000(2)), an international organization   that is dedicated to promoting the industry of intelligent agents by openly developing specifications supporting interoperability among agents and   agent-based applications. A standard process for the CNP involves four basic   steps as shown in Figure 2

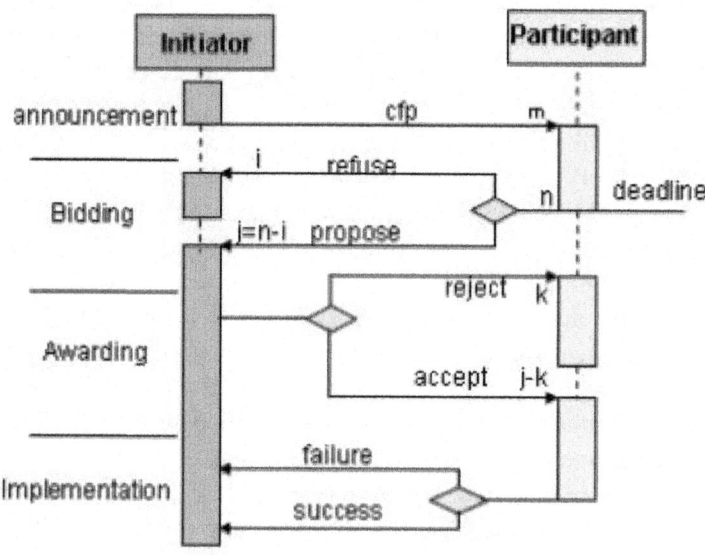

Figure 2: FIPA Contract Net Protocol (FIPA 2000(1))

- Task announ-cement: The initiator agent broadcasts an announcement to the par-ticipant agents to call for proposal (cfp).
- Bidding: Those participants that receive the announcement and have the appropriate capability to make the evaluation on the task, and then reply their bids to the initiator agent.
- Awarding: The initiator agent awards the task to the most appropriate agents according to the proposals they have submitted.
- Implementation: The awarded participant agent performs the task, and re-ceives the benefits predefined.

Currently, the CNP method is widely used for negotiation and coordination among agent systems, and has been proved to be effective to solve distributed problems. The development platform for agent-based systems Most of the intelligent agent and multi-agent systems are working under dis- tributed and heterogeneous environments, and C++ and Java are the two most- adopted programming languages. At the early stage, some works were devel- oped from scratch, which were rather difficult to deal with. Recently, useful platforms and templates have been provided by some institutes, which can provide some basic and necessary modules such as communication, interface design, agent kernel template, etc. The adoption of these platforms facilitates the development and let the designers focus on the functional modules programming, thus to reduce the workload and difficulty of agent applications development. Among these development platforms, JADE (F. Bellifemine, Caire et al. 2006) and Jatlite (JATLite) are two typical and widely applied sys- tems.   JADE (Java Agent DEvelopment Framework) is a software framework devel- oped in Java language by TILAB (JADE 2005). It is composed of two parts, one   is the libraries (Java classes) required to develop the agent applications and   functions and the other is a run-time environment providing some necessary   services for the agents' execution. The platform can be executed in a distrib- uted, multi-party application with peer-to-peer communication, which include   both wired and wireless environment. The platform supports execution with   cross operation system and the configuration can be controlled via a remote   GUI; furthermore, the platform also supports hot exchange, moving agents   from one machine to another at run-time.   In JADE, middleware acts as the interface of low layer and applications. Each agent is identified by a unique name and provides a set of services. The agent can search for other agents to provide given services according to the middleware if necessary. With the role of middleware, the agents can dynamically discover other agents and to communicate with them by a peer-to-peer paradigm. The structure of a message complies with the ACL language defined by FIPA and includes fields, such as variables indicating the context a message re- fers-to and timeout that can be waited before an answer is received, aiming at   supporting complex interactions and multiple parallel conversations (F. Bellifemine, Caire et al. 2006). Furthermore, in order to support the implementation of complex conversations, JADE provides a set of skeletons of typical in- teraction patterns to perform specific tasks, such as negotiations, auctions and   task delegation.   Compared with JADE, JATLite is a lighter and easier to use as a platform for   agent-based applications. As it provides only some basic and necessary func- tions for agent applications, JATLite is more suitable for prototype develop- ment in agent-based research work. JATLite is composed of some java pack- ages which help to build agent-based applications with Java language. In the   package, four different layers: abstract, base, KQML and router layer, covering   from the lowest layer with an operation system to the router function. The   package is developed according to TCP/IP protocols, which ensures the system   can be running in the Internet. In JATLite, the router acts as the key role in the   message communication among the agents. Although the functions of JATLite may not be as powerful as those in JADE, it   is still widely used. The platform is simple and provides some reliable basic services for the agent execution. Furthermore, it sill provides some templates for agent execution; thus, the designers can implement their applications easily.

Application of MAS in manufacturing system integration

With manufacturing systems become distributed and decentralized in different geographical sites, it is necessary to study the solution of specific problems which arise in a distributed environment. As the MAS system shows the promising capability to solve distributed problems, a great amount of efforts have been made to apply the multi-agent theory to the manufacturing system integration, aiming to study the problems of the distributed manufacturing system. In this part, some typical agent-based manufacturing systems are introduced.

MetaMorph

MetaMorph and MetaMorph II are two consecutive projects developed in the University of Calgary (Shen, Maturana et al. 1998; Shen, Xue et al. 1998; Maturana, Shen et al. 1999; Shen 2002). MetaMorph is an adaptive multi-agent manufacturing system aimed to provide an agent-based approach for dynamically creating and managing agent communities in distributed manufacturing environments (Maturana, Shen et al. 1999). There are two main types of agents in MetaMorph: resource agents and mediator agents. Resource agents are used to   represent manufacturing devices and operations, and mediator agents are   used to coordinate the interactions among agents (resource agents and also   mediator agents).   Mediator-centric federation architecture is one of the system characteristics, by   which the intelligent agents can link with mediator agents to find other agents   in the environment. The activity for mediators is interpreting messages, de- composing tasks, and providing processing times for every new task. Addi- tionally, mediators assume the role of system coordinators by promoting co- operation among the intelligent agents. Both brokering and recruiting   communication are adopted to find the related agents for specific tasks. Once   appropriate agents have been found, these agents can be directly linked and   communicate directly without the aid of mediator. The object of MetaMorph II project is to integrate the manufacturing enterprise's activities such as design, planning, scheduling, and simulation, execution, with those of its suppliers, customers and partners into a distributed intelligent open environment. In this Infrastructure, the manufacturing system is primarily organized at the highest level through 'subsystem' mediators. Each subsystem is connected (integrated) to the system through a special mediator. Each subsystem itself can be an agent-based system (e.g., agent-based manufacturing scheduling system), or any other type of system like the feature-based design system, knowledge-based material management system, and so on. Agents in a subsystem may also be autonomous agents at the subsystem level. Some of these agents may also be able to communicate directly with other subsystems or the agents in other subsystems. Mediators are also agents, called mediator agents. The main difference between a mediator and a facilitator is that a facilitator provides the message services in general, but a mediator assumes an additional role of system coordinators by promoting cooperation among intelligent agents and learning from the agents' behavior.

CIIMPLEX

CIIMPLEX (Consortium for Intelligent Integrated Manufacturing Planning-Execution) (Peng, Finin et al. 1998) was developed by UMBC and some other institutes, which presents an agent-based framework of enterprise integration for manufacturing planning and execution. The system is composed of name server, facilitator agent and gateway agent and some executive agents. The different functions of the manufacturing process are encapsulated into individual agents. In the system, a set of agents with specialized expertise can be quickly assembled to gather the relevant information and knowledge, and to cooperate with other agents to arrive at timely decisions to deal with various enterprise scenarios.   Different executive agents are designed to perform special functions such as   data collection, analysis of plans and schedules, resolving the conflicts; fur- thermore, some agents are created to integrate the function of the legacy sys- tem. With this architecture, the raw transaction data of the low level, such as   shop floors activities, can be collected, aggregated, interpolated and extrapo- lated by agents and made available for other interested agents. Manufacturing   planning and execution can thus be integrated through the collaboration of   these agents.

The AARIA project

The AARIA project (Autonomous Agents at Rock Island Arsenal) (Parunak, Baker et al. 1998; Parunak, Savit et al. 1998) is an agent-based prototype system based on the Internet-related technologies. In the system, Internet is used as the platform, and distributed scheduling and controlling techniques are developed to realize the distributed manufacturing. All of the agents are tied by Internet to form a virtual manufacturing environment for tasks. With the agent technology, the resource can be redeployed easily to meet the fast changing environment, which increases the agility of the system. Furthermore, the productive resources can be adjusted according to the products' requirement, which make the system meet the customization requirements.   In the system, besides the functional decomposition, physical factors are also   considered during the resource agentificaiton process. The main agents of the   system include resource brokers, part brokers, and unit process brokers. Re- source broker agents manage the constrained resources of the system (e.g.   people, machines, facilities, etc.). Part broker agents manage material handling   and inventory. Unit process broker agents utilize their knowledge of how to combine resources and parts to make other parts. These three types of agents negotiate among themselves and with the customer along the axes of possible production including price, quality, delivery time, product features, and speed of answers (Baker, Parunak et al. 1999).

DaimlerChrysler manufacturing line control system One industrial application of agent-based manufacturing line control system is implemented in DaimlerChrysler (Bussmann and Schild 2001; Bussmann and Sieverding 2001), whose objective is to develop a flexible transportation and control system. In this project, each work piece, machine and shifting table is encapsulated into one specific agent. In the execution, the work piece agent will auction off its coming operations to machine agents. Every machine agent's bid include information about its current state of buffer. Once a work piece agent awards a machine agent, it will be the next goal of the work piece. The routing of the work piece will be negotiated by the work piece agent with the shifting tale agent. The application of agent-based system shows two key advantages for product manufacturing. One is the distributed responsiveness, as the decision making can be much more localized. If unexpected events occur, agents have the autonomy and proactiveness to try alternatives thus can be more responsive to prevailing circumstances. The other advantage is that dynamical control mechanism, which improves the agility of the system. Because the schedules are built up dynamically through flexible interactions, they can be readily altered in the event of delays or unexpected contingencies. The implementation of the testing system has increased throughput and greater robustness to failure (Jennings and Bussmann 2003), which also shows a good prospect for the agent-based manufacturing system.

Multi-Agent Based Distributed Product Design and Manufacturing

Planning In this section, one agent-based distributed manufacturing system developed in the National University of Singapore (NUS) (Sun 1999; Jia 2001; Wang 2001; Jia, Fuh et al. 2002; Li 2002; Jia, Ong et al. 2004; Mahesh, Fuh et al. 2005) is presented, which studies a multi-agent based approach to integrate product design, manufacturability analysis, process planning and scheduling in a distributed manner. Under this framework, geographically dispersed entities are allowed to work cooperatively towards overall manufacturing system goals. The system model considers constraints and requirements from the different product development cycles and manufacturing.

The system adopted a federator structure to model the various manufacturing functional departments in a manufacturing process that includes design, manufacturability evaluation, process planning, scheduling and shop floor control. In the system, the different functional departments dispersed in different geographical sites are encapsulated into agents. Facilitator architecture is selected as the system architecture, which comprises a facilitator agent, a console agent and several service agents. The facilitator is responsible for the decomposition and dispatch of tasks, and resolving conflicts of system execution. The console agent acts as an interacting interface between designers and the system. The service agent models the functional modules of different product development phases, including Designing Agent, Manufacturing Resource Agent, Manufacturability Evaluation Agent, Process Planning Agent, Scheduling Agent, etc. Each functional agent represents a participant involved in a different product development and manufacturing phase. Facilitator plays the central and control roles in the whole environment, and each participant can know the status and requirements of other participants in real-time through it.

System framework design

In a multi-agent manufacturing environment, the isolated and distributed functional sub-systems can be integrated by encapsulating them as interacting agents. Each agent is specifically in charge of a particular design or manufacturing activity. The agents communicate and exchange information to solve problems in a collaborative manner. The components interact dynamically, addressing the different manufacturing planning issues collaboratively, thereby avoiding costly manual iterations. The federated structure adopted as the architecture ensures the openness of the system, which makes the functional agents can join or leave without having to halt or to reinitialize the other agents' work in progress. The different components can interact dynamically in such platform, addressing the product design and manufacturing planning issues efficiently, and the separate domains of expertise may reside at distributed sites on a network but collaborate with others on a common task, which results in great time saving in terms of data transfer and interpretation. Some legacy software tools can also be wrapped into Java-based agents having the capability of interacting with others.

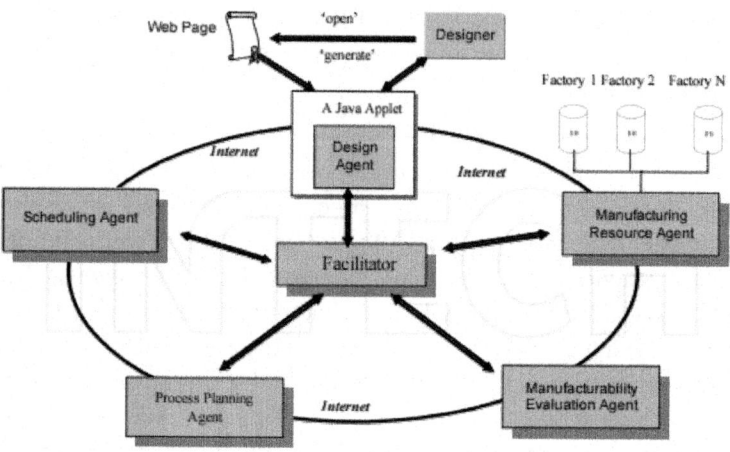

Figure 3: System architecture.

The system architecture, as depicted in figure 3, which is composed of six components: Facilitator, Design Agent (D-Agent), Manufacturing Resource Agent (MR-Agent), Manufacturability Evaluation Agent (ME-Agent), Process Planning Agent (PP-Agent) and Scheduling Agent (S-Agent). The last four service agents (encapsulated pre-existing legacy tools) and the D-Agent interact with each other through the Facilitator.

Agent coordination and individual agents

The function of each agent during this framework is defined as follows:

Facilitator

It is responsible for the management of interactions and conflict resolution in the agent community. Once any agent joins or leaves the system, it needs to register with status and information to the Facilitator. Thus, the Facilitator "knows" which agent is available, and any function each agent has. Each executive agent receives tasks from the facilitator, and feedbacks the results to it after completing. The Facilitator also routes the requests information received to appropriate agents based on its knowledge of capabilities of each agent, which is known as content-based routing. In performing this task, the Facilita- tor can go beyond a simple pattern matching by translating messages, decomposing problems into sub-problems, and scheduling the work on those subproblems.

D-Agent

It is the interface between the system and the designers, by which the design information of product is submitted to other agents for manufacturability analysis, process plan and scheduling generation. Once the designed parts need further modifications, the information will be also sent back. It also advises the designer to make necessary modifications to the design.

MR-Agent:

This agent manages manufacturing resource models from those different factories of the system, which contain information of available shop-floor resources, including machines and tools, and the capability of these resources. These models are stored in individual databases located at different local sites. The agent is in charge of looking for a suitable capability for manufacturability evaluation.

ME-Agent:

This agent is responsible for the manufacturability evaluation of the product design with the help of acquiring capability information from the MR-Agent. It returns information about conflicts to the Facilitator, as well as suggestions for product redesign or a suitable capability model.

PP-Agent:

This agent is responsible for the generation of an optimal process plan based on the design and selected resources.

S-Agent:

This agent makes the manufacturing scheduling for parts, and feedback to the facilitator. In order to manage the product and manufacturing information, each agent has a local database, which is used to store and manage messages received from other agents.

Furthermore, with the Internet, all of these individual databases are integrated into a distributed database to improve the execution efficiency of the system.

Under such framework, the manufacturing tasks are usually executed by the cooperation of several different related agents. The takes are decomposed firstly into some sub-tasks and dispatched to the destination for process, which needs the cooperation and coordination of the agents in the system. In the project, the agents of the system make negotiations trying to find optimal tradeoffs among their local preferences and other agents' preferences and make commitments based on the negotiation results. The task-completing process in the system consists of the following steps:(1)A remote designer submits design information of a product/part to the Facilitator via the D-Agent; (2) The Facilitator decomposes the task into mutually interrelated sub-tasks from a global objective point of view; (3)The Facilitator dispatches the sub-tasks to appropriate executive agents; (4)Executive agents complete their sub-tasks independently; (5)The Facilitator detects conflicts;(6)The Facilitator defines and refines the shared space of interacting agents to remove conflicts; and (7) Conflict resolution.

AGENT DEFINITION

Manufacturability Evaluation Agent (MEA)

This agent is in charge of evaluating the manufacturability of a designed part during the design phrase and sending the modifying information to the design agent if necessary. The agent judges the manufacturability for one part and selects the most preferable machining plan alternatives considering the part's dimensions, tolerances, and surface finishes, along with the availability and capabilities of machine tools and tooling constraints. MEA is, firstly, to check whether the design features are defined correctly; secondly to check if design features can be machined or not based on the current available manufacturing resources; and thirdly to find out all available manufacturing resources that can fabricate the product. Manufacturability Evaluation After receiving the feature information from the Facilitator, the ME-Agent carries out a manufacturability evaluation process for the design. It starts with a local manufacturability evaluation on the model in terms of design flaws. Any local conflict detected in the process is notified to the D-Agent by the Facilitator for design modification. Upon the completion of local manufacturability evaluations, the ME-Agent makes a global manufacturability evaluation on the model by acquiring a factory model from the RC-Agent.

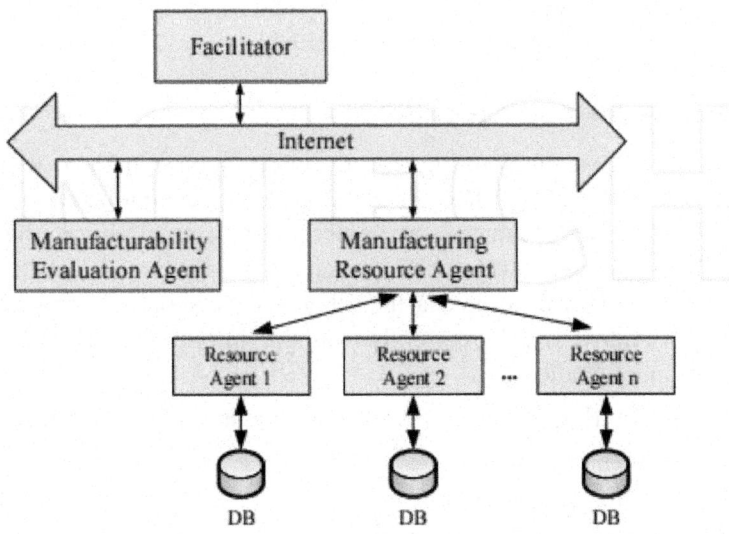

Figure 4: Manufacturability evaluation modules.

The RC-Agent checks the availability of various resources required to create the part. Any global conflict is notified to the MR-Agent for it to search for a suitable factory model as a substitute to the former. The two analysis processes are repeatedly executed until no conflict is found on the part model and the agent analyses four manufacturability issues as follows:

(1)Design flaws:

The design flaws refer to those features which are difficult or impossible to machine. The ME-Agent identifies all possible flaws to avoid much higher rectification cost at an advanced stage.

(2)Tool accessibility:

The ME-Agent checks the tool accessibility of each feature. A feature may be inaccessible due to its location and orientation. For those flaws, the cutting tool may not work correctly and need to be modified. (3)Availability of cutters: The ME-Agent checks whether all the required cutting tools to machine the part are available in the factory under consideration. If some machined features exceed the manufacturing capability of the cutting tools available, then, they will need further revision on the design.

Tolerance and surface finish requirements: The ME-Agent also need to check the capability of machines contained in the factory against the tolerance and surface finish requirements in the part model.

Resource Coordination Agent (RCA)

The RCA collects the manufacturing resource information from the work shops and factories with the help of RA. As the system is open and heterogeneous, it is needed that the agent can support flexibility and extensibility in dynamic manufacturing environments, which means that resource coordination, including interaction with users via the other agents, should be sensitive to both the query context and the currently available information. The RCA has the following functionality:

For a manufacturing resource request, multiple instantiations of the search node may be created;

- Task execution is data-driven;
- Parse the query, and decompose it if appropriate parsing involves getting an ontological model;
- Construct KIF queries based on the SQL queries' contents, and query the Resource Agent using the KIF queries to find relevant resources.

Resource Agent

The Resource Agent (RA) manages the data contained in manufacturing information source (e.g., distributed systems database) available to retrieve and update. It acts as an interface between the local data source and other agents, hiding details of the local data organization and representation. To accomplish this task, an RA can announce and update its presence, location and the description of its contents to the RCA. There are two types of information that is of potential interest to other agents: 1. value (ranges) of chosen data objects, 2. the set of operations allowed on the data. The operations range from a single read/update to more complicated data analysis operations. The advertisement information can be sent to the RCA.

RA also needs to answer queries from other agents. It has to translate queries expressing in a common query language (such as KQML) into a language understood by the underlying system. This translation is facilitated by a mapping between the local data concepts and terms, as well as between the common query language syntax, semantics and operators, and those of the native language. Once the queries are translated, the RA sends them to the manufacturing resource database for execution, and translates the answers back into the format understood by the RCA. Additionally, RA and the underlying data source may group certain operations requested by other agents into a local transaction. In addition, RA provides limited transaction capabilities for global resource transaction.

Process planning agent

Process planning agent is developed to generate the optimal or near-optimal process plans for designed part based on the criterion chosen. Under the distributed environment, factories possessing various machines and tools are dispersed at different geographical locations, and usually different manufacturing capabilities are selected to achieve the highest production efficiency. When jobs requiring several operations are received, feasible process plans are produced by available factories according to the precedence relationships of the operations. The final optimal or near-optimal process plan will emerge after comparison of all the feasible process plans. In order to realize and optimize the process plan for the distributed manufacturing systems, the Genetic Algorithm (GA) methodology is adopted as an optimizing method. The GA method is composed of four operations as following: encoding, population initialization, reproduction, and chromosome evaluation and selection. Encoding When dealing with a distributed manufacturing system, a chromosome not only needs to represent the sequence of the operations but also indicate which factory this process plan comes from. Therefore, the identity number of the factory will be placed as the first gene of each chromosome no matter how the other genes are randomly arranged. Each other gene comprises the operation ID and corresponding machine, tool and tool access direction (TAD), which will be used to accomplish the operation. As a result, a process plan including factory and operation information will be represented by a random combination of genes. Population Initialization The generation of the initial population in GA is usually done randomly; however, the initial population must consist of strings of valid sequences, satisfying all precedence relations. Once the number of initialized chromosomes is prescribed, the procedures of initialization are given as follows: (1) Randomly select one factory ID number from the available factory list. (2) Randomly select one operation among those, which have no predecessors. (3) Among the remaining operations, randomly select one which has no predecessor or which either predecessor all have already been selected. (4) Repeat step (3) until each operation has been selected for only once. (5) Revisit the first selected operation. (6) Randomly select machines and tools from the selected factory that can be used for performing the operation. (7) Randomly select one amongst all possible TADs for the operation. (8) Repeat steps (6) and (7), until each operation has been assigned a machine, tool and TAD. (9) Repeat steps (1) to (8) until the number of prescribed chromosome is reached.

Reproduction

A genetic search starts with a randomly generated initial population; further generations are created by applying GA operators. This eventually leads to a generation of high performing individuals. There are usually three operators in a typical genetic algorithm, namely crossover operator, mutation operator and inversion operator. In the proposed GA, mutation and crossover operators are used for gene recombination, which is also called offspring generation.

Crossover

In this step, a crossover operator is adopted to ensure the local precedence of operations is met and a feasible offspring is generated. The procedure of the crossover operation is described as follows: (1) Randomly choose two chromosomes as parent chromosomes. (2) Based on the chromosome length, two crossover points are randomly gen erated to select a segment in one parent. Each string is then divided into three parts, the left side, the middle segment and the right side according to the cutting points. (3) Copy the left side and right side of parent 1 to form the left side and right side of child 1. According to the order of operations in parent 2, the opera tor constructs the middle segment of child 1 with operations of parent 2, whose IDs are the same as operations of the middle segment in parent 1. (4) The role of these parents will then be exchanged in order to generate an other offspring child 2. (5) Re-assign machines and tools to the operations in the middle segment to legalize the offspring chromosomes according to the factory id.

Mutation

Mutation operator is used to investigate some of the unvisited points in the search space and also to avoid pre-mature convergence of the entire feasible space caused by some super chromosomes. A typical GA mutation makes changes by simply exchanging the positions of some randomly selected genes. However, for the distributed manufacturing system, mutation once is not enough to explore all the feasible operation sequences, as well as compare the different selected factory combination. In the proposed GA process, mutation happens to the chromosomes twice, one is for selected factory (mutation 1) and the other is for the operations (mutation 2).

The procedure of mutation 1 is described as follows: (1) Randomly select a factory ID from the factory ID list, which is different from the current one.

(2) In order to legalize the chromosome, machines and tools will be re- as signed for all the operations according to the new factory-id. The procedure of mutation 2 is depicted as follows: (1) Randomly choose a chromosome. (2) Choose several pairs of genes stochastically and permute their positions.

Chromosome

Evaluation When all the individuals (process plans) in the population have been determined to be feasible, i.e. an operation precedence is guaranteed, they can be evaluated based on the objective functions. The objective of the CAPP problem is to obtain an optimal operation sequence that results in optimizing resources and minimizing production costs as well as processing time. In this research, two optimization criteria, i.e. minimum processing times and minimum production cost, are employed to calculate the fitness of each process plan and measure the efficiency of a manufacturing system. After the completion of the manufacturability evaluation, the PP-Agent generates an optimal process plan for the factory supplied by the ME-Agent. The agent first constructs the solution space by identifying all the possible operation-methods (OpM's) for machining each feature and then uses a GA to search for the best plan according to a specific criterion. The criterion can be constructed by using the following cost factors: Machine cost (MC)

$$MC = \sum_{i=1}^{n} MCI_i$$

(1)

where n is the total number of OpM's and MCI_i is the machine cost index for using machine-i, a constant for a particular machine.

   Tool cost (TC)

$$TC = \sum_{i=1}^{n} TCI_i$$

(2)

where TCI_i is the tool cost index for using tool-i, a constant for a particular tool. Machine change cost (MCC): a machine change is needed when two adjacent operations are performed on different machines.

$$MCC = MCCI \times \sum_{i}^{n-1} \Omega(M_{i+1} - M_i)$$

(3)

where MCCI is the machine change cost index, a constant and M_i is the ID of the machine used for operation i. Setup change cost (SCC): a setup change is needed when two adjacent OpM's performed on the same machine have different Tool Approaching Directions (TADs).

$$SCC = SCCI \times \sum_{i=1}^{n-1} ((1 - \Omega(M_{i+1} - M_i)) \times \Omega(TAD_{i+1} - TAD_i))$$

(4)

where SCCI is the setup change cost index, a constant. Tool change cost (TCC): a tool change is needed when two adjacent OpM's performed on the same machine use different tools

$$TCC = TCCI \times \sum_{i=1}^{n-1} ((1 - \Omega(M_{i+1} - M_i)) \times \Omega(T_{i+1} - T_i))$$

(5)

where TCCI is the tool change cost index, a constant.

Scheduling Agent

In a distributed manufacturing environment, every factory has its particular niche areas and can outperform other factories in those specific aspects; therefore, one batch of products are to be finished by the most suitable factory combination with the considerations of low cost and short make span. To meet such requirement, each available candidate factory will submit a feasible process plan for a product in the batch it is capable of processing. The agent then compares all the candidate factories, selects the final factory combination for the products, and meanwhile arranges the manufacturing operations in an optimal sequence. In brief, to generate an optimal schedule in a distributed manufacturing environment, there are two determining factors: the selected factory (or process plan) for every product and operations' sequencing of the machines in the factories. Here, GA is also used as the optimization method to achieve better scheduling results. The scheduling agent is mainly composed of four major components: the scheduling kernel including the GA engine, the stand-alone scheduling module, scheduling agent module, and the e-scheduling module. Among these four parts, the scheduling kernel can be categorized as the basic component, while the stand-alone module, scheduling agent module and e-scheduling module can be categorized as the application components. The basic component could be combined with any of the three application components to form a scheduling entity, which can carry out the scheduling tasks solely based on a specified scheduling objective.

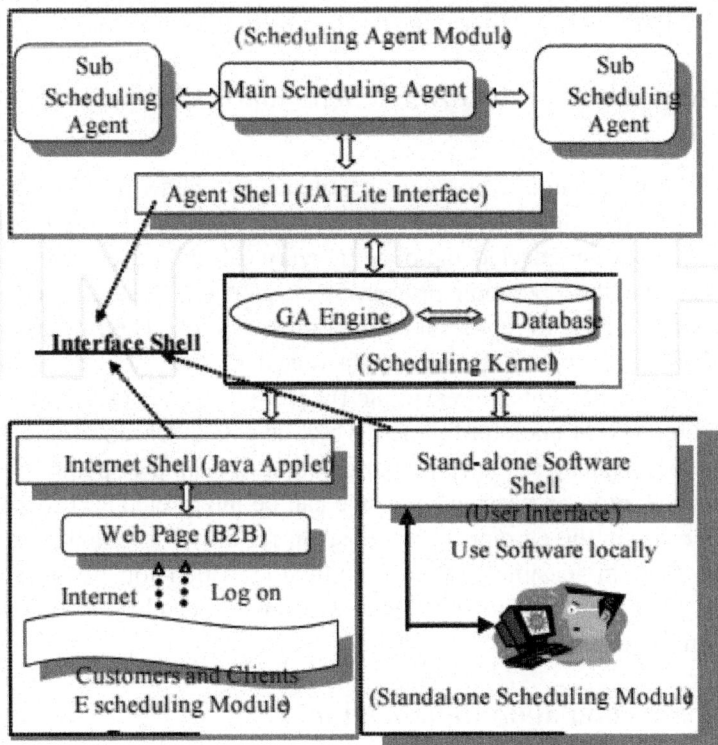

Figure 5: Scheduling agent.

Integrating with the scheduling kernel, each of the three application modules has a particular working mode. The scheduling agent module includes one main scheduling agent (MSA) and many distributed sub-scheduling agents (SSAs). The scheduling agent structure consists of not only the parallel sub-scheduling agents (SSAs) but also a main scheduling agent (MSA). Such a structure can effectively facilitate the information communication among the different production participants within the distributed scheduling system. The MSA, used by the production management department, is responsible for regulating the job production operations, collecting the factories' information from the SSAs and then, making scheduling through the scheduling kernel. After the scheduling, the details about which job is manufactured in which factory in what time and what machine is responsible for what operation can be obtained and sent to each SSA. To finish the job production cooperatively, the SSAs are distributed in the manufacturing factories, each representing a factory, collecting the factory's working status, detecting its dispatched jobs, and checking the manufacturing progress. In addition, if any contingency happens in any factory, the SSA will send the accident information to the MSA, wait and execute the re-arranged schedule made by MSA.

Genetic Algorithm for distributed Scheduling Chromosome representation

To handle the distributed scheduling problems, the genes in the GA must comprise the two dominant factors in the distributed manufacturing environment, i.e., the selected factory for every job with the corresponding process plan and the operation processing sequence. Here, a four-element string is used as a gene to represent an operation of a job. The first element is the identification number of the selected factory that is used to process the job, and the next three elements represent the ID of a job. Thus, a schedule can be represented by a combination of genes, which is called "chromosome" in the GA terminology, as long as the combination comprises all the operations of the jobs. Every operation processing sequence can be interpreted according to its occurrence order in the chromosome. As such, for any distributed scheduling problem, a random feasible schedule, including which job goes to which factory and what is the operation processing sequence, can be encoded using a combination of genes.

Chromosome Population Initialization

To begin the search for the best chromosome and correspondingly, the factory combination and the optimal schedule the chromosome represents, a number of chromosomes are initialized as follows: (1) Create the lists of ID number of the feasible factories for every job. If job 'j03' can be processed in factories '1', '3' and '5', then the list for job 'j03' will be 1, 3, and 5. (2) For every job, randomly select a factory ID number from the job's feasible factory list. (3) According to the job's process plan in the selected factory, produce the job's operation genes. For example: the genes, 1j03-1j03-1j03, mean the first, second and third (last) operation of job 'j03'. All the operations will be manufactured in factory '1'. (4) Repeat step (2) and step (3), until there is no job left. (5) Combine and mix all the produced genes together in a stochastic way to form a chromosome. (6) Repeat step (5), until a prescribed number of chromosome populations is formed. Genetic operators The power and potential of GA method come from the gene recombination, including crossover, mutation and inversion, which explore all the possible search space. In this proposed GA, two genetic operators, crossover and mutation, are employed for the gene recombination, which is called offspring generation. The procedures for the crossover operation are described as follows: (1) Choose two chromosomes as parents and exchange a random partial string (genes) in the two parents to generate two offspring chromosomes. (2) Regulate (delete or compensate) genes in each offspring chromosome so that it comprises the operations of all the jobs and inherits the genetic traits of their parents. Because the sequence of the genes in the chromosome expresses the jobs' operation processing sequence, after the crossover, the processing sequence for every operation (gene) in the schedule (chromosome) is changed. It is important to note that the precedence of a job's operations will not be affected by the crossover because every gene (operation) of a job in the chromosome is not fixed to a specific operation of the job and it is interpreted according to the order of occurrence in the sequence for a given chromosome. The crossover is carried out under the assumption that a factory has been selected for every job. Yet, in the distributed manufacturing environment, the randomly selected factory is, by and large, not necessarily the most suitable one for the specific job. Therefore, another genetic operator, gene mutation, is employed twice for modifying the operation processing sequence again as well as changing the selected candidate factory for the jobs. The mutation procedures are as follows: (1) In a chromosome, choose several pairs of genes stochastically and permute their positions (Mutation1). (2) Select one gene (job) from the chromosome in a random manner. In the meantime, randomly select a factory ID number from the job's feasible factory list, which has been created in step (1) of chromosome population initialization. (3) In the selected chromosome, replace the first element (represents the ID of the selected factories) of all the genes (the gene selected in step 2) with the newly selected factory ID number (Mutation2). The aim of the step is to change the factory selection used to manufacture the job. (4) To keep the consistency of the chromosomes, apply the same change of factory selection for the gene (job) to all the other chromosomes in their generation. In step (1), the first mutation changes the jobs' operation sequences, while from step (2) to step (4), the second mutation changes the factory ID, which is used for a randomly selected job. After gene crossover and mutation, the parent chromosomes can produce a generation of offspring. In the same way, the offspring could reproduce the next generation of offspring. Thus, through this iteration, numerous chromosomes are produced and can be compared. Correspondingly during this process, many possible factory combinations and job operation processing sequences are formed and analyzed. The application of such gene crossover and mutation in this GA ensures each product (job) to be manufactured in its most suitable factory. In addition, the production schedule of each of the factories in the distributed environment can be generated concurrently.

Prototype Implementation

In order to verify the effectiveness of the system, a prototype of the proposed system has been developed for the integration of design, manufacturability analysis, and process planning. The developed system includes a unique facilitator, and several functional agents, which are organized according to the framework of Figure 3. JATLite is selected as the template for the agents' development. Each agent is composed of the following components: network interface, local knowledge model and domain knowledge model. All the agents in the system use the common communication protocol, KQML, for concurrent negotiations. KQML is conceived as both a message format and a message-handling protocol to support run-time knowledge sharing among agents. It is essentially a wrapper to encode context sensitive information. The KQML language is di- vided into three layers: the content layer, the message layer, and the communication layer, as shown in figure 6.

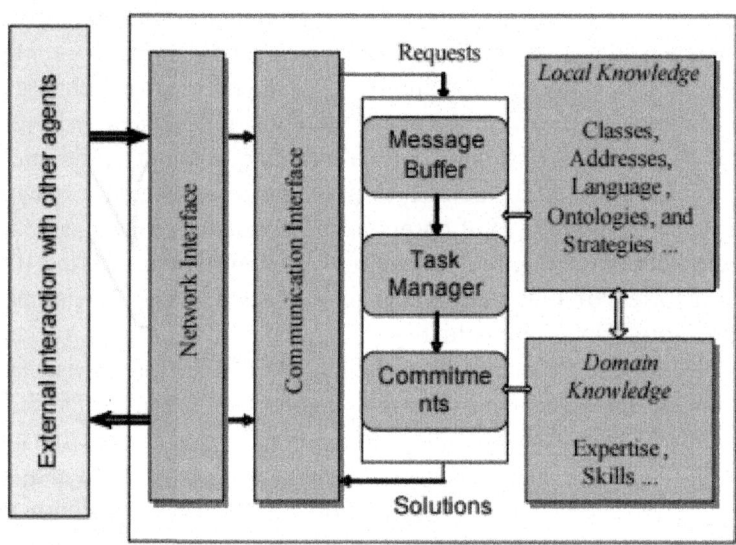

Figure 6: Internal structure of an agent

In the system, agents are autonomous cognitive entities, with deductive, storage and communication capabilities. Autonomy in this case means that an agent can function independently from any other agent. There are three kinds of functional agent in the system. Each has different internal structure, and can be decomposed into the following components: (1) A network interface: It couples the agent to the network. (2) A communication interface: It is composed of several methods or functions for communicating with other agents. (3) A local function module: Resources in this model include Java classes to perform the desired functions, other agent names, messaging types in KQML syntax.

The functional module also provides the facility of inference and collaboration facilities. The collaboration facilities are the general plans for coordination behaviour that balances the agent's local interests with global (community) interests. (4) Agent Knowledge base: The model comprises expertise that is required for an agent to perform functional tasks, and skills that may be methods for ac tivating actions corresponding to the received requests. A snap shot on the prototype system (PPA and MSA) is shown in figure 7 below.

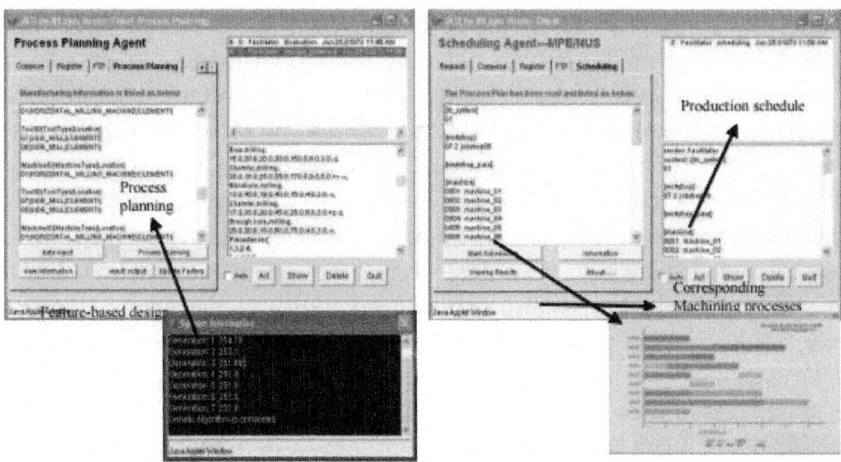

Figure 7: Optimal process plans and production schedules generated from the PPA and MSA respectively

CONCLUSION AND FUTURE WORK

Agent theory was developed from the distributed artificial intelligence around 20 years before. As its nature and characteristic are suitable for distribute problems solving, agent theory has been viewed as a promising theory and methodology applied to a distributed environment. It has now achieved some promising results in industrial applications. The rapid development of Internet has also provided a good tool and suitable platform for agent's application. With the use of Internet and advancement of communication methods, agents can be dispersed in different geographical places, which make it easy for collaboration of different partners in a manufacturing supply chain. Except the description to some basic theories and key issues in agent-based systems and the introduction to some typical manufacturing systems, this chapter also introduces one agent-based distributed manufacturing system developed in NUS. The objective of this research is to develop a distributed collaborative design environment for supporting cooperation among existing engineering tools organized as independent agents on different platforms. A facilitator multi-agent system architecture is discussed for distributed collaborative product design and manufacturing planning, and one prototype for collaborative design and planning on machining processes, which has been developed as a proof-of-concept system, demonstrates the capability of such a multi-agent approach to integrate design, manufacturability evaluation, process planning and scheduling efficiently. This approach could be extended to include other product life cycle considerations, collaboratively, at the design stage. Some of the advantage of the system over the traditional mode can be achieved in several facets as follows: 1) Distributed function execution improves the efficiency and the reliability of the system, thus to increase the responsiveness of the enterprise to the market requirements. In the system, the major functional tasks are distrib uted, and each one agent needs to focus on one task execution, which can improve the efficiency of the process. Furthermore, as the functions are executed dispersedly, once some functional agent malfunction, the rest can still work, which also improves the reliability of the whole system. 2) Open architecture to make the system good adaptability and easy exten sion. As the system adopts a facilitator structure, the newly added agent can execute the system tasks if only registering with the facilitator; at the same time the functional agent can leave the system only need to inform the facilitator and not influence other's progress. Moreover, the newly functional agents can be also integrated into the system without interfering other agents' function. All of these can make the system good adaptability and easy extension, thus to improve the agility of the system. 3) The agent-based system provides a platform to realize the concurrent func tion in the product development. The different departments from design through manufacture to customer service can join together for one product design, thus to improve the quality and efficiency of the final product de velopment. Although some promising results have been achieved in the prototype and previous research work on the agent theory, there are still difficulties to be overcome for its wider application in industry. Some of the challenges faced include: 1) Effective coordination and negotiation methods for MAS. Coordination and negation methods are the basis and key issues for intelligent agent sys tems. It has been under study for a long time and a variety of methods have been proposed, but one effective and efficient method for the agent-based system is still needed. 2) Methods to incorporate and agentify the legacy manufacturing systems and tools. Now, there are various computer-aided software systems ap plied in manufacturing system and industrial scenario, but there is still no successful methodology to agentify these legacy modules in the agent sys tems. This is one bottleneck that impedes a wider development of agent- based methods for industrial applications. 3). Agent theory provides a decentralized solution for complex systems, de composing and conquering make the agents easy to deal with subtasks. But one problem emerges is that the local optimization can not result in a global optimal result for the whole system. How to achieve a global opti mal solution in the agent-based system still needs a further study

REFERENCES

1.   Baker, A. D., H. V. D. Parunak, et al. (1999). Internet-based Manufacturing: A Perspective from the AARIA Project, Enterprise Action Group.

2.   Bussmann, S. and K. Schild (2001). An agent-based approach to the control of flexible production systems. ETFA 2001. The 8th International Conference on Emerging Technologies and Factory Automation. Proceedings, 15-18 Oct. 2001, Antibes-Juan les Pins, France, IEEE.

3.   Bussmann, S. and J. Sieverding (2001). Holonic control of an engine assembly plant: an industrial evaluation. Proceedings of IEEE International Conference on Systems, Man & Cybernetics, 7-10 Oct. 2001, Tucson, AZ, USA, IEEE.

4.   Castelfranchi, C. (1995). Guarantees for autonomy in cognitive agent architecture. Proceedings of the workshop on agent theories, architectures, and languages on Intelligent agents, Amsterdam, the Netherlands, SpringerVerlag New York, Inc.

5.   Davis, R. and R. G. Smith (1983). "Negotiation as a metaphor for distributed problem solving." Artificial Intelligence 20(1): 63-109. F. Bellifemine, G. Caire, et al. (2006). JADE: A White Paper. FIPA (1997). FIPA 97 Part 2 Version 2.0: Agent Communication Language Specification, FIPA.

6.   FIPA (2000(1)). FIPA Interaction Protocol Library Specification, FIPA. FIPA (2000(2)). FIPA ACL Message Structure Specification, FIPA. Gensereth, M. R. and S. P. Ketchpel (1994). "Software Agents." Communications of the ACM Vol. 37(No. 7): 48-53.

7.   JADE (2005). JADE:Java Agent DEvelopment Framework, http://jade. tilab. com/. JATLite http://java.stanford.edu/

8.   Jennings, N. R. and S. Bussmann (2003). "Agent-based control systems: Why are they suited to engineering complex systems?" IEEE Control Systems Magazine 23(3): 61-73.

9.   Jennings, N. R. and M. Wooldridge (1998). Applications of intelligent agents, Springer-Verlag New York, Inc.

10. Jia, H. Z.,2001, Internet-based multi-functional scheduling for

distributed manufacturing systems, M. Eng Thesis ,National University of Singapore,Singapore.

11. Jia, H. Z., J. Y. H. Fuh, et al. (2002). "Web-based Multi-functional Scheduling System for a Distributed Manufacturing Environment." Concurrent Engineering 10(1): 27-39.

12. Jia, H. Z., S. K. Ong, et al. (2004). "An adaptive and upgradable agent-based system for coordinated product development and manufacture." Robotics and Computer-Integrated Manufacturing 20(2): 79-90

13. Li, L.,2002, Agent-based computer-aided process planning for distributed manufacturing systems., M. Eng Thesis, National University of Singapore,Singapore.

14. Mahesh, M., J.Y.H.Fuh, et al. (2005). "Towards A Generic Distributed and Collaborative Digital Manufacturing", Proceedings of the International Manufacturing Leaders Forum on Global Competitive Manufacturing, Adelaide, Australia.

15. Maturana, F., W. Shen, et al. (1999). "MetaMorph: an adaptive agent-based architecture for intelligent manufacturing." International Journal of Production Research 37(10): 2159-73.

16. Parunak, H. V. D., A. D. Baker, et al. (1998). "The AARIA Agent Architecture: from Manufacturing Requirements to Agent-Based System Design". WorkshopProc. on Agent-Based Manufacturing, ICAA'98, Minneapolis, MN.

17. Parunak, H. V. D., R. Savit, et al. (1998). Agent-Based Modeling vs. EquationBased Modeling: A Case Study and Users' Guide. Proceedings of the First International Workshop on Multi-Agent Systems and Agent-Based Simulation, Springer-Verlag, London, UK.

18. Peng, Y., T. Finin, et al. (1998). "A Multi-Agent System for Enterprise Integration." International Journal of Agile Manufacturing, vol. 1(No. 2): 201-212.

19. Shen, W. (2002). "Distributed manufacturing scheduling using intelligent agents." Intelligent Systems, IEEE [see also IEEE Intelligent Systems and Their Applications] 17(1): 88-94.

20. Shen, W., F. Maturana, et al. (1998). Learning in Agent-Based Manufacturing Systems. Proceedings of AI & Manufacturing Research Planning Workshop, IAlbuquerque, NM, The AAAI Press,.

21. Shen, W., D. Xue, et al. (1998). An Agent-Based Manufacturing Enterprise Infrastructure for Distributed Integrated Intelligent Manufacturing Systems.

22. Proceedings of the 3rd International Conference on the Practical Applications of Agents and Multi-Agent Systems, London, UK. Smith, R. G. (1980). "

23. The contract net protocol: high level communication and control in a distributed problem solver." IEEE Transactions on Computers C-29(12): 1104-1113. Smith, R. G. (1988).

24. The contract net protocol: high-level communication and control in a distributed problem solver Distributed Artificial Intelligence Morgan Kaufmann Publishers Inc.: 357-366 Sun, J.,1999,

25. Agent-based product design and planning for distributed concurrent engineering, M.

26. Eng Thesis, National University of Singapore,Singapore. Wang, G.,2001,

27. Agent-based manufactuirng service system, M. Eng. Thesis, National University of Singapore. Weiss, G. (1999).

28. Muliagent Systems: A Modern Approach to Distributed Artificial Intelligence. Cambridge, Massachusetts, The MIT Press.

29. Wooldridge, M. and N. R. Jennings (1995). "Intelligent Agents: Theories and Practices." Knowledge Engineering Review: 115-152.

Chapter 9

USING OVERALL EQUIPMENT EFFECTIVENESS FOR MANUFACTURING SYSTEM DESIGN

Vittorio Cesarotti[1], Alessio Giuiusa[1, 2] and Vito Introna[1]

[1]University of Rome "Tor Vergata", Italy

[2]Area Manager Inbound Operations at Amazon.com

INTRODUCTION

Different metrics for measuring and analyzing the productivity of manufacturing systems have been studied for several decades. The traditional metrics for measuring productivity were *throughput* and *utilization rate*, which only measure part of the performance of manufacturing equipment. But, they were not very helpful for *"identifying the problems and underlying improvements needed to increase productivity"* [1].

During the last years, several societal elements have raised the interest in analyze the phenomena underlying the identification of productive performance parameters as: capacity, production throughput, utilization, saturation, availability, quality, etc.

This rising interest has highlighted the need for more rigorously defined and acknowledged productivity metrics that allow to take into account a set of synthetic but important factors (availability, performance and quality) [1]. Most relevant causes identified in literature are:

- The growing attention devoted by the management to cost reduction approaches [2] [3];
- The interest connected to successful eastern productions approaches, like Total *Productive Maintenance* [4], *World Class Manufacturing* [5] or *Lean production* [6];
- The importance to go beyond the limits of traditional business management control system [7];

For this reasons, a variety of new performance concepts have been developed. The total productive maintenance (TPM) concept, launched by Seiichi Nakajima [4] in the 1980s, has provided probably the most acknowledged and widespread quantitative metric for the measure of the productivity of any production equipment in a factory: the *Overall Equipment Effectiveness* (OEE). OEE is an appropriate measure for manufacturing organizations and it has being used broadly in manufacturing industry, typically to monitor and control the performance (time losses) of an equipment/work station within a production system [8]. The OEE allows to quantify and to assign all the time losses, that affect an equipment whilst the production, to three standard categories. Being standard and widely acknowledged, OEE has constituted a powerful tool for production systems performance benchmarking and characterization, as also the starting point for several analysis techniques, continuous improvement and research [9] [10]. Despite this widespread and relevance, the use of OEE presents limitations. As a matter of fact, OEE focus is on the single equipment, yet the performance of a single equipment in a production system is generally influenced by the performance of other systems to which it is interconnected. The time losses propagation from a station to another may widely affect the performance of a single equipment. Since OEE measures the performance of the equipment within the specific system, a low value of OEE for a given equipment can depend either on little performance of the equipment itself and/or time losses propagation due to other interconnected equipments of the system.

This issue has been widely investigated in literature through the introduction of a new metric: the Overall Equipment Effectiveness (OTE), that considers the whole production system as a whole. OTE embraces the performance losses of a production system both due to the equipments and their interactions.

Process Designers need usually to identify the number of each equipments necessary to realize each activity of the production process, considering the interaction and consequent time losses a priori. Hence, for a proper design of the system, we believe that the OEE provides designer with better information on each equipment than OTE. In this chapter we will show how OEE can be used to carry out a correct equipments sizing and an effective production system design, taking into account both equipment time losses and their propagation throughout the whole production system.

In the first paragraph we will show the approach that a process designer should face when designing a new production system starting from scratch.

In the second paragraph we will investigate the typical time-losses that affect a production system, although are independent from the production system itself.

In the third part we will define all the internal time losses that need to be considered when assessing the OEE, along with the description of a set of critical factors related to OEE assessment, such as buffer-sizing and choice of the plant layout.

In the fourth paragraph we will show and quantify how time losses of a single equipment affects the whole system and vice-versa.

Finally, we will show through the simulation some real cases in which a process design have been fully completed, considering both equipment and time losses propagation.

MANUFACTURING SYSTEM DESIGN: ESTABLISH THE NUMBER OF PRODUCTION MACHINES

Each process designer, when starting the design of a new production system, must ensure that the number of equipments necessary to carry out a given process activity (e.g. metal milling) is sufficient to realize the required volume. Still, the designer must generally ensure that the minimum number of equipment is bought due to elevated investment costs. Clearly, the performance inefficiencies and their propagation became critical, when the purchase of an extra (set of) equipment(s) is required to offset time losses propagation. From a price strategy perspective, the process designer is generally requested to assure the number of requested equipments is effectively the minimum possible for the requested volume. Any not necessary over-sizing results in an extra investment cost for the company, compromising the economical performance.

Typically, the general equation to assess the number of equipments needed to process a demand of products (D) within a total calendar time C_t (usually one year) can be written as follow (1):

$$n_i = int\left[\frac{D^*ct_i}{C_t^*\vartheta^*\eta_i}\right] + 1$$

(1)

Where:

- D is the number of products that must be produced;
- cticti is theoretical cycle time for the equipment I to process a piece of product;
- ct_i is the number of hours (or minutes) in one year.
- ϑ is a coefficient that includes all the external time losses that affect a production system, precluding production.
- η_i I is the efficiency of the equipment I within the system.

It is therefore possible to define L_t, Loading time, as the percentage of total calendar time C_t that is actually scheduled for operation (2):

$$L_t = C_t^*\vartheta$$

(2)

The equation (1) shows that the process designer must consider in his/her analysis three parameters unknown a priori, which influence dramatically the production system sizing and play a key role in the design of the system in order to realize the desired throughput. These parameters affect the total time available for production and the real time each equipment request to realize a piece [9], and are respectively:

- External time losses, which are considered in the analysis with ϑ;
- The theoretical time cycle, which depends upon the selected equipment(s);
- The efficiency of the equipment which depends upon the selected equipments and their interactions, in accordance to the specific design.

This list highlights the complexity implicitly involved in a process design. Several forecasts and assumptions may be required. In this sense, it is a good practice to ensure that the ratio in equation (3)is always respected for each equipment:

$$\frac{\left(\frac{D^*ct_i}{L_t{}^*\eta_i}\right)}{n_i} < 1 \tag{3}$$

As a good practice, to ensure (3) being properly lower than 1 allows to embrace, among others, the variability and uncertainty implicitly embedded within the demand forecast. In the next paragraph we will analyze the External time losses that must be considered during the design.

EXTERNAL TIME LOSSES

Background

For the design of a production system several time-losses, of different nature, need to be considered. Literature is plenty of classifications in this sense, although they can diverge one each others in parameters, number, categorization, level of detail, etc. [11] [12]. Usually each classification is tailored on a set of sensible drivers, such as data availability, expected results, etc. [13]. One relevant classification of both external and internal time losses is provided by Grando et al. [14]. Starting from this classification and focusing on external time losses only, we will briefly introduce a description of common time-losses in Operations Management, highlighting which are most relevant and which are negligible under certain hypothesis for the design of a production system (Table 1).

The categories LT1 and LT2 don't affect the performance of a single equipment, nor influence the propagation of time-losses throughout the production system.

Still, it is important to notice that some causes, even though labeled as external, are complex to asses during the design. Despite these causes are external, and well known by operations manager, due to the implicit complexity in assessing them, these are detected only when the production system is working via the OEE, with consequence on OEE values. For example, the lack of material feeding a production line does not depend by the OEE of the specific station/equipment. Nevertheless when lack of material occurs a station cannot produce with consequences on equipment efficiency, detected by the OEE. (4).

Table 1: Adapted from Grando et al. 2005

Symbol	Name	Description	Synonyms
Lt1	Idle times resulting from law regulations or corporate decisions	Summer vacations, holidays, shifts, special events (earthquakes, flood);	System External Causes
Lt2	Unplanned time	Lack of demand; Lack of material in stocks;	System External Causes
		Lack of orders in the production plan;	
		Lack of energy; Lack of manpower (strikes, absenteeism);	
		Technical tests and manufacturing of nonmarketable products; Training of workers;	
Lt3	Stand by time	Micro-absenteism, shift changes; physiological increases; man machine interaction;	Machine External Causes; System External Causes
		Lack of raw material stocks for single machines; Unsuitable physical and chemical properties of the available material;	
		Lack of service vehicle; Failure to other machines;	

Considerations

The external time losses assessment may vary in accordance to theirs categories, historical available data and other exogenous factors. Some stops are established for through internal policies (e.g. number of shift, production system closure for holidays, etc.). Other macro-stops are assessed (e.g. Opening time to satisfy forecasted demand), whereas others are considered as a forfeit in accordance to the Operations Manager Experience. It is not possible to provide a general magnitude order because, the extent of time losses depend from a variety of characteristic factor connected mainly to the specific process and the specific firm. Among the most common ways to assess this time losses we found: Historical data, Benchmarking with similar production system, Operations Manager Experience, Corporate Policies.

The Calendar time C_t is reduced after the external time losses. The percentage of C_t in which the production system does not produce is expressed by $(1- \vartheta)$, affecting consequently the L_t (2).

These parameters should be considered carefully by system designers in assessing the loading time (2). Although these parameters do not propagate throughout the line their consideration is fundamental to ensure the identification of a proper number of equipments.

Idle Times

There is a set of idle times that result from law regulations or corporate decisions. These stops are generally known a-priori, since they are regulated by local law and usually contribute to the production plant localization-decision process. Only causes external to the production system are responsible for their presence.

Unplanned Times

The unplanned time are generally generated by system external causes connected with machineries, production planning and production risks.

A whole production system (e.g. production line) or some equipment may be temporarily used for non marketable product (e.g. prototype), or they may are not supposed to produce, due to test (e.g. for law requirements), installation of new equipments and the related activities (e.g. training of workers).

Similarly, a production system may face idle time because of lack of demand, absence of a production schedule (ineffectiveness of marketing function or production planning activities) or lack of material in stock due to ineffectiveness in managing the orders. Clearly, the presence of a production schedule in a production system is independent by the Operations manager and by the production system design as well. Yet, the lack of stock material, although independent from the production system design is one of the critical responsibility of any OM (inventory management).

Among this set of time losses we find also other external factors that affect the system availability, which are usually managed by companies as a risk. In this sense occurrence of phenomenon like the lack of energy or the presence of strikes are risks that companies well know and that usually manage according to one of the four risk management strategy (avoidance, transfer, mitigation acceptance) depending on their impact and probability.

Stand By Time

Finally, the stand-by time losses are a set of losses due to system internal causes, but still equipment external causes. This time losses may affect widely the OTE of the production line and depend on: work organization losses, raw material and material handling.

Micro-absenteeism and shift changes may affect the performances of all the system that are based on man-machine interaction, such as the production equipments or the transportation systems as well. Lack of performance may propagate throughout the whole system as other equipment ineffectiveness. Even so, Operations manager can't avoid these losses by designing a better production line. Effective strategies in this sense are connected with social science that aim to achieve the employee engagement in the workplace [15].

Nonetheless Operations Manager can avoid the physiological increases by choosing ergonomic workstations.

The production system can present other time-losses because of the raw material, both in term of lack and quality:

- Lack of raw material causes the interruption of the throughput. Since we have already considered the ineffective management of the orders in "Unplanned Time", the other related causes of time-losses depend on demand fluctuation or in ineffectiveness of the suppliers as well. In both cases the presence of safety stock allows operations manager to reduce or eliminate theirs effects.

- Low raw material standard quality (e.g. physical and chemical properties), may affect dramatically the performance of the system. Production resource (time, equipment, etc) are used to elaborate a throughput without value (or with a lower value) because of little raw material quality. Also in this case, this time losses do not affect the design of a production system, under the hypothesis that Operations Manager ensures the raw material quality is respected (e.g. incoming goods inspection). The missed detection of low quality raw materials can lead the Operations Manager to attribute the cause of defectiveness to the equipment (or set of equipment) where the defect is detected.

Considering the Vehicle based internal transport, a broader set of considerations is requested. Given two consecutive stations *i-j*, the vehicles make available the output of station i to station j (figure 1).

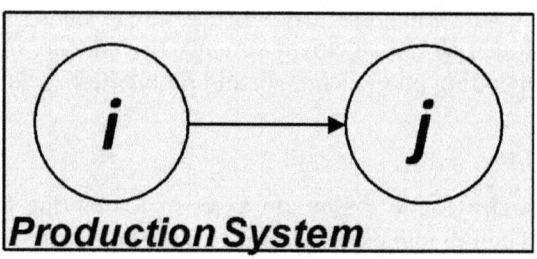

Figure 1: Vehicle based internal transport: transport the output of station i to the station j.

In this sense any vehicle can be considered as an equipment that is carrying out the transformation on a piece, moving the piece itself from station i to station j (Figure 2).

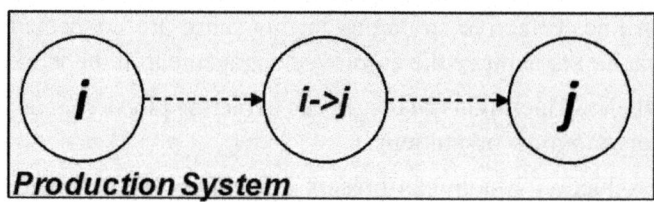

Figure 2: Service vehicles that connect i-j can be represented as a station itself amid i-j.

The activity to transport the output from station i to station j is a transformation (position) itself. Like the equipments, also the service vehicles affect and are affected by the OTE. In this sense successive considerations on equipments losses categorization, OEE, and their propagations throughout the system, OTE, can be extended to service vehicles. Hence, the design of service vehicles would be carried out according to the same guidelines we provide in successive section of this chapter.

THE FORMULATION OF OEE

In this paragraph we will provide process designer with a set of topics that need to be addressed when considering the OEE during the design of a new production system. A proper assessment a-priori of the OEE, and the consequent design and sizing of the system demand process designer to consider a variety of complex factors, all related with OEE. It is important to notice that OEE measures not only the internal losses of efficiency, but is also detects time losses due to external time losses (par.2.1, par.2.2). Hence, in this paragraph we will firstly define analytically the OEE. Secondly we will investigate, through the analysis of relevant literature, the relation between the OEE of single equipment and the OEE of the production system as a set of interconnected equipments. Then we will describe how different time losses categories, of an equipment, affect both the OEE of the equipment and the OEE of the Whole system. Finally we will debate how OEE need to be considered with different perspective in accordance to factors as ways to realize the production and plant layout.

Mathematical Formulation

OEE is formulated as a function of a number of mutually exclusive components, such as *availability efficiency*, *performance efficiency*, and *quality efficiency* in order to quantify various types of productivity losses.

OEE is a value variable from 0 to 100%. An high value of OEE indicates that machine is operating close to its maximum efficiency. Although the OEE does not diagnose a specific reason why a machine is not running as efficiently as possible, it does give some insight into the reason [16]. It is therefore possible to analyze these areas to determine where the lack of efficiency is occurring: breakdown, set-up and adjustment, idling and minor storage, reduced speed, and quality defect and rework [1] [4].

In literature exist a meaningful set of time losses classification related to the three reported efficiencies (availability, performance and quality). Grando et al. [14] for example provided a meaningful and comprehensive classification of the time-losses that affect a single equipment, considering its interaction in the interaction system. Waters et al. [9] and Chase et al. [17] showed a variety of acknowledged possible efficiency losses schemes, while Nakajima [4] defined the most acknowledged classification of the "6 big losses".

In accordance with Nakajima notations, the conventional formula for OEE can be written as follow [1]:

$$OEE = A_{eff} \, Pe_{eff} \, Q_{eff} \tag{4}$$

$$A_{eff} = \frac{T_u}{T_t} \tag{5}$$

$$Pe_{eff} = \frac{T_p}{T_u} * \frac{R_{avg}^{(a)}}{R_{avg}^{(th)}} \tag{6}$$

$$Q_{eff} = \frac{P_g}{P_a} \tag{7}$$

Table 2 summarizes briefly each factor.

Table 2: OEE factors description

Factor	Description
A_{eff}	Availability efficiency. It considers failure and maintenance downtime and time devoted to indirect production task (e.g. set up, changeovers).
Pe_{eff}	Performance efficiency. It consider minor stoppages and time losses caused by speed reduction
Q_{eff}	Quality efficiency. It consider loss of production caused by scraps and rework.
T_u	Equipment uptime during the T_t. It is lower that T_t because of failure, maintenance and set up.
T_t	Total time of observation.
T_p	Equipment production time. It is lower than T_t because of minor stoppages, resets, adjustments following changeovers.
$R_{avg}^{(a)}$	Average actual processing rate for equipment in production for actual product output. It is lower than theoretical ($R_{avg}^{(th)}$) because of speed/production rate slowdowns.
$R_{avg}^{(th)}$	Average theoretical processing rate for actual product output.
P_g	Good product output from equipment during T_t.
P_a	Actual product units processed by equipment during T_t. We assume that for each product rework the same cycle time is requested.

The OEE analysis, if based on single equipment data, is not sufficient, since *no machine is isolated in a factory, but operates in a linked and complex environment* [18]. A set of inter-dependent relations between two or more equipments of a production system generally exists, which leads to the propagation of availability, performance and quality losses throughout the system.

Mutual influence between two consecutive stations occurs even if both stations are working ideally. In fact if two consecutive stations (e.g. station A and station B) present different cycle times, the faster station (eg. Station A = 100 pcs/hour) need to reduce/stop its production rate in accordance with the other station production rate (e.g. Station B = 80 pcs/hour).

Station A	Station B
100 pcs/hour	80 pcs/hour

In this case, the detected OEE of station A would be 80%, even if any efficiency loss occurs. This losses propagation is due to the unbalanced cycle time.

Therefore, when considering the OEE of equipment in a given manufacturing system, the measured OEE is always the performance of the equipment within the specific system. This leads to practical consequence for the design of the system itself.

A comprehensive analysis of the production system performance can be reached by extending the concept of OEE, as the performance of individual equipment, up to factory level [18]. In this sense OEE metric is well accepted as an effective measure of manufacturing performance not only for single machine but also for the whole production system [19] and it is known as *Overall Throughput Effectiveness* OTE [1] [20].

We refer to OTE as the OEE of the whole production system.

Therefore we can talk of:

- Equipment OEE, as the OEE of the single equipment, which measures the performance of the equipment in the given production system.

- System OEE (or OTE), which is the performance of the whole system and can be defined as the performance of the bottleneck equipment in the given production system.

An Analytical Formulation to Study Equipment and System OEE

$$\text{System OEE} = \frac{\text{Number of good parts produced by system in total time}}{\text{Theoretical number of parts produced by system in total time}}$$

(8)

The System OEE measures the systemic performance of a manufacturing system (productive line, floor, factory) which combines activities, relationships between different machines and processes, integrating information, decisions and actions across many independents systems and subsystem [1]. For its optimization it is necessary to improve coordinately many interdependent activities. This will also increase the focus on the plant-wide picture.

Figure 3 clarify which is the difference between Equipment OEE and System OEE, showing how the performance of each equipment affects and is affected by the performances of the other connected equipments. These time losses propagation result on a Overall System OEE. Considering the figure 3 we can indeed argue that given a set of $i=1,..,n$ equipments, OEE_i of the ith equipment depends on the process in which it has been introduced, due to the availability, performance and quality losses propagation.

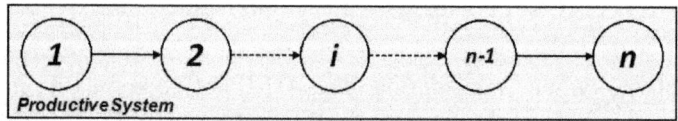

Figure 3: A production system composed of n stations.

According to the model proposed by Huang et al in [1], the System OEE (OTE) for a series of n connected subsystems, is formulated in function of theoretical production rate $R_{avg(F)}^{(th)}$ relating to the slowest machine (the bottleneck), theoretical production rate $R_{avg}^{(th)}(N)$ and OEE_0 of nth station as shown in (9):

$$OTE = \frac{OEE_n \times R_{avg(n)}^{th}}{R_{avg(F)}^{th}}$$
(9)

The OEE_n computed in (9) is the OEE of nth station introduced in the production system (the OEE_n when n is in the system and it is influenced by the performance of other $n-1$ equipments).

According to (9) the only measure of OEE_n is a measure of the performance of the whole system (OTE). This is true because performance data on n are gathered when the station n is already working in the system with the other $n-1$ station and, therefore, its performance is affected from the performance of the other $n-1$ prior stations. This means that the model proposed by Huang, could be used *only when the system exists and it is running,* so OEE_n could be directly measured on field.

But during system design, when only technical data of single equipment are known, the same formulation in (9) can't be used, since without information on the system OEE_n in unknown a-priori. Hence, in this case the (9) couldn't provide a correct value of OTE.

How Equipment Time-Losses Influence the System Performance and Vice-Versa

The OEE of each equipment, *as isolated machine* (independent by other station) is affected only by (5),(6) and (7) theoretical intrinsic value. But once the equipment is part of a system its performance depends also upon the interaction with other *n-1* equipments and thus on their performance. It is now more evident why, for a correct estimate and/or analysis of equipment OEE and system OEE, it is necessary to take into account losses propagation. These differences between single subsystem and entire system need to be deeply analyzed to understand real causes of system efficiency looses. In particular their investigation is fundamental during the design process, because a correct evaluation of OEE and for the study of effective losses reduction actions (i.e. buffer capacity dimensioning, quality control station positioning); but also during the normal execution of the operations because it leads to correct evaluation of causes of efficiency losses and their real impact on the system.

The table 3 shows how efficiency losses of a single subsystem (e.g. an equipment/ machine), given by Nakajima [4] can spread to other subsystem (e.g. in series machines) and then to whole system.

In accordance to table 3 a relevant lack of coordination in deploying available factory resources (people, information, materials, and tools) by using OEE metric (based on single equipment) exists. Hence, a wider approach for a holistic production system design has to focus also *on the performance of the whole factory* [18], resulting by the interactions of its equipments.

Table 3: Example of propagation of losses in the system

	Single subsystem	Entire system
Availability	Breakdown losses Set-up and adjust-ment	Downtimes losses of upstream unit could slackening production rate of downstream unit without fair buffer capacity Downtimes losses of downstream unit could slackening production rate of upstream unit without fair buffer capacity
Perfor-mance	Idling and minor stoppages Reduced speed	Minor stoppages and speed reduction could influencing production rate of the downstream and upstream unit in absence of buffer
Quality	Quality defects and rework Yield losses	Production scraps and rework are losses for entire process depends on where the scraps are identified, rejected or reworked in the process

This issue have been widely debated and acknowledged in literature [1] [18]. Several Authors [8] [21] have recognized and analyzed the need for a coherent, systematic methodology for design at the factory level.

Furthermore, the following activities, according to [18] [21] have to be considered as OTE is also started at the factory design level:

- Quality (better equipment reliability, higher yields, less rework, no misprocessing);
- Agility and responsiveness (more customization, fast response to unexpected changes, simpler integration);
- Technology changes;
- Speed (faster ramp up, shorter cycle times, faster delivery);
- Production cost (better asset utilization, higher throughput, less inventory, less setup, less idle time);

At present, there is not a common well defined and proven methodology for the analysis of System OEE [1] [19] *during the system design.* By the way the effect of efficiency losses propagation must be considered and deeply analyzed to understand and eliminate the causes before the production system is realized. In this sense the *simulation* is considered the most reliable method, to date, in designing, studying and analyzing the manufacturing systems and its dynamic performance [1] [19]. Discrete event simulation and advanced process control are the most representatives of such areas [22].

Layout Impact on OEE

Finally, it is important to consider how the focus of the design may vary according the type of production system. In flow-shop production system the design mostly focuses on the OTE of the whole production line, whereas in job-shop production system the analysis may focus either on the OEE of a single equipment or in those of the specific shop floor, rather than those of the whole production system. This is due to the intrinsic factors that underlies a layout configuration choice.

Flow shop production systems are typical of high volume and low variety production. The equipment present all a similar cycle time [23] and is usually organized in a product layout where interoperation buffers are small or absent. Due to similarity among the equipments that compose the production system, the saturation level of the different equipments are likely to be similar one each other. The OEE are similar as well. In this sense the focus of the analysis will be on loss time propagation causes, with the aim to avoid their occurrence to rise the OTE of the system.

On the other hand, in *job shop production systems*, due to the specific nature of operations (multi-flows, different productive paths, need for process flexibility rather than efficiency) characterized by higher idle time and higher stand-by-time, lower values of performances index are pursued.

Different products categories usually require a different sequence of tasks within the same production system so the equipment is organized in a process layout. In this case rather than focusing on efficiency, the design focuses on production system flexibility and in the layout optimization in order to ensure that different production processes can take place effectively.

Generally different processes, to produce different products, imply that bottleneck may shift from a station to another due to different production processes and different processing time of each station in accordance to the specific processed product as well.

Due to the shift of bottleneck the presence of buffers between the stations usually allows different stations to work in an asynchronous manner, consecutively reducing/eliminating the propagation of low utilization rates.

Nevertheless, when the productive mix is known and stable over time, the study of plant layout can embrace bottleneck optimization for each product of the mix, since a lower flexibility is demanded.

The analysis of quality propagation amid two or more stations should not be a relevant issue in job shop, since defects are usually detected and managed within the specific station.

Still, in several manufacturing system, despite a flow shop production, the equipment is organized in a process layout due to physical attributes of equipment (e.g. manufacturing of electrical cables showed in § 4) or different operational condition (e.g. pharmaceutical sector). In this case usually buffers are present and their size can dramatically influence the OTE of the production system.

In an explicit attempt to avoid unmanageable models, we will now provide process designers and operations managers with useful hints and suggestion about the effect of inefficiencies propagation among a production line along with the development of a set of simulation scenarios (§ 3.5).

OEE AND OTE Factors for Production System Design

OEE is formulated as a function of a number of mutually exclusive components, such as availability efficiency, performance efficiency, and quality efficiency in order to quantify various types of productivity losses.

During the design of the production system the use of intrinsic performance index for the sizing of each equipment although wrong could seem the only rational approach for the design. By the way, this approach don't consider the interaction between the stations. Someone can argue that to make independent each station from the other stations through the buffer would simplify the design and increase the availability. Still, the interposition of a buffer between two or more station may not be possible for several reason. Most relevant are:

- logistic (space unavailability, huge size of the product, compact plant layout, etc.);
- economic (the creation of stock amid each couple of station increase the WIP and consequently interest on current assets);
- performance;
- product features (buffer increase cross times, critical for perishable products);

In our model we will show how a production system can be defined considering availability, performance and quality efficiency (5),(6), (7) of each station along with their interactions. The method embraces a set of hints and suggestions (best practices) that lead designers in handle interactions and losses propagation with the aim to rise the expected performance of the system. Furthermore, through the development of a simulation model of a real production system for the electrical cable production we provide students with a clear understanding of how time-losses propagate in a real manufacturing system.

The design process of a new production system should always include the simulation of the identified solution, since the simulation provides designer with a holistic understanding of the system. In this sense in this paragraph we provide a method where the design of a production system is an iterative process: the simulation output is the input of a successive design step, until the designed system meet the expected performance and performance are validated by simulation. Each loss will be firstly described referring to a single equipment, than its effect will be analyzed considering the whole system, also throughout the support of simulation tools.

Set Up Availability

Availability losses due to set up and changeover must be considered during the design of the plant. In accordance with the production mix, the number of set-up generally results as a trade-off between the set up costs (due to loss of availability + substituted tools, etc.) and the warehouse cost.

During the design phase some relevant consideration connected with set-up time losses should be considered. A production line is composed of n stations. The same line can usually produce more than one product type. Depending on the difference between different product types a changeover in one or more stations of the line can be required. Usually, the more negligible the differences between the products, the lower the number of equipments subjected to set up (e.g. it is sufficient the set up only of the label machine to change the labels of a product depending on the destination country). In a given line of n equipments, if a set up is requested in station i, loss availability can interest only the single equipment I or the whole production line, depending on the buffer presence, their location and dimension:

- If buffers are not present, the set up of station i implies the stop of the whole line (figure 4). This is a typical configuration of flow shop process realized by one or more production line as food, beverages, pharmaceutical packaging,....

- If buffers are present (before and beyond the station i) and their size is sufficient to decouple the station i by the other i-1 and i+1 station during the whole set up, the line continues to work regularly (figure 5).

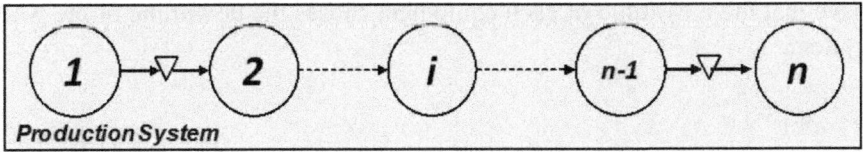

Figure 4: Barely decoupled/Coupled Production System (buffer unimportant or null).

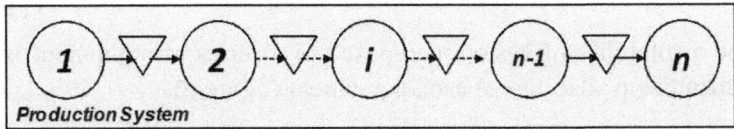

Figure 5: Decoupled Production System.

Hence, the buffer design plays a key role in the phenomena of losses propagation throughout the line not only for set-up losses, but also for other availability losses and performance losses as well. The degree of propagation ranges according to the buffer size amid zero (total dependence-maximum propagation) and maximum buffer size (total independence-no propagation). It will be debated in the following (§ 3.5.3), when considering the performance losses, although the same principles can be applied to avoid propagation of minor set up losses (mostly for short set-up/changeover, like adjustment and calibrations).

Maintenance Availability

The availability of an equipment [24] is defined as $A_{eff} = \dfrac{T_u}{T_i}$. The availability of the whole production system can be defined similarly. Nevertheless it depends upon the equipment configurations. Operations Manager, through the choice of equipment configurations can increase the maintenance availability. This is a design decision, since different equipments must be bought and installed according to desired availability level. The choice of the configuration usually results as a trade-off between equipment costs and system availability. The two main equipment configuration (not-redundant system, redundant system) are debated in the following.

Not redundant system

When a system is composed of non redundant equipment, each station produces only if the equipment is working.

Hence if we consider a line of n equipment connected a s a series we have that the downtime of each equipment causes the downtime of the whole system.

$$A_{system} = \prod_{i=1}^{n} A_i$$

(10)

$$A_{system} = \prod_{i=1}^{n} A_i = 0,\ 7^*0,\ 8^*0,\ 9 = 0,\ 504$$

(11)

The availability of system composed of a series of equipment is always lower than the availability of each equipment (figure 6).

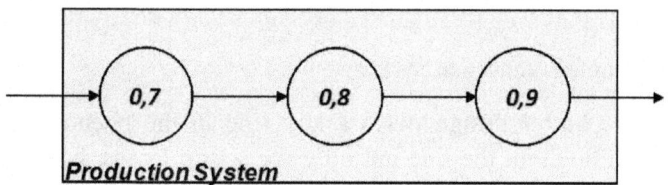

Figure 6: Availability of not redundant System.

Total redundant system

Oppositely, to avoid failure propagation amid stations, designer can set the line with a total redundancy of a given equipment. In this case only the contemporaneous downtime of both equipments causes the downtime of the whole system.

$$A_{system} = 1 - \prod_{i=1}^{n} (1 - A_i)$$

(12)

In the example in figure 7 we have two single equipments connected with a redundant system of two equipment (dotted line system).

Hence, the redundant system availability (dotted line system) rises from 0,8 (of the single equipment) up to:

$$A_{parallel} = 1 - \prod_{i=1}^{n}(1 - A_i) = \left(1 - 0, 8\right)*\left(1 - 0, 8\right) = 0, 96$$

(13)

Consequently the availability of the whole system will be:

$$A_{system} = \prod_{i=1}^{n} A_i = 0, 7*\left[0, 96\right]*0, 9 = 0, 6048$$

(14)

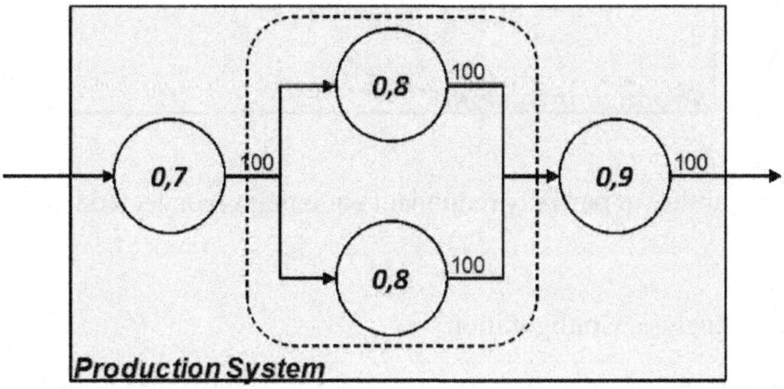

Figure 7: Availability of totally redundant equipments connected with not redundant equipments

To achieve an higher level of availability it has been necessary to buy two identical equipments (dou.ble cost). Hence, the higher value of availability of the system should be worth economically.

Partial redundancy

An intermediate solution can be the partial redundancy of an equipment. This is named K/n system, where n is the total number of equipment of the parallel system and k is the minimum number of the nequipment that must work properly to ensure the throughput is produced. The figure 8 shows an example.

The capacity of equipment b', b'' and b''' is 50 pieces in the referral time unit. If the three systems must ensure a throughput of 100 pieces, it is at least necessary that $k=2$ of the $n=3$ equipment produce 50 pieces. The table 4 shows the configuration states which ensure the output is produced and the relative probability that each state manifests.

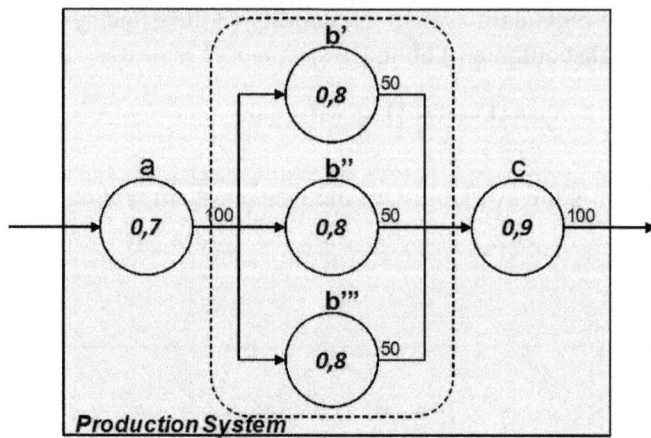

Figure 8: Availability of partially redundant equipments connected with not redundant equipments.

Table 4: State Analysis Configuration

b'	b''	b'''	Probability of occurrance	[*100]
UP	UP	UP	0,8*0,8*0,8	0,512
UP	UP	DOWN	0,8*0,8*(1-0,8)	0,128
UP	DOWN	UP	0,8*(1-0,8)*0,8	0,128
DOWN	UP	UP	(1-0,8)*0,8*0,8	0,128
Total Avail-ability				0,896

In this example all equipments b have the same reliability ($0,8$), hence the probability the system of three equipment ensure the output should have been calculated, without the state analysis configuration (table 4), through the binomial distribution:

$$R_{k/n} = \sum_{j=k}^{n} \binom{n}{j} R^{j}[1-R]^{n-j}$$

(15)

$$R_{2/3} = \binom{3}{2} 0,8^2[1-0,8] + \binom{3}{3} 0,8^3 = 0,896$$

(16)

Hence, the availability of the system (a, b'-b''-b''', c) will be:

$$A_{system} = \prod_{i=1}^{n} A_i = 0,\ 7*\left[0,\ 896\right]*0,\ 9 = 0,\ 56448$$

$$(17)$$

In this case the investment in redundancy is lower than the previous. It is clear how the choice of the level of availability is a trade-off between fix-cost (due to equipment investment) and lack of availability.

In all the cases we considered the buffer as null.

When reliability of the equipments (b in our example) the binomial distribution (16) is not applicable, therefore the state analysis configuration (table 4) is required.

Redundancy with Modular Capacity

Another configuration is possible.

The production system can be designed as composed of two equipment which singular capacity is lower than the requested but which sum is higher. In this case if it is possible to modulate the production capacity of previous and successive stations the expected throughput will be higher than the output of a singular equipment.

Considering the example in figure 9 when b' and b'' are both up the throughput of the subsystem b'-b'' is 100, since capacity of a and c is 100. Supposing that capacity of a and c is modular, when b' is down the subsystem can produce 60 pieces in the time unit. Similarly, when b'' is down the subsystem can produce 70. Hence, the expected amount of pieces produced by b'-b'' is 84,8 pieces (table 5). When considering the whole system if either a or c are down the system cannot produce. Hence, the expected throughput in the considered time unit must be reduced of the availability of the two equipments:

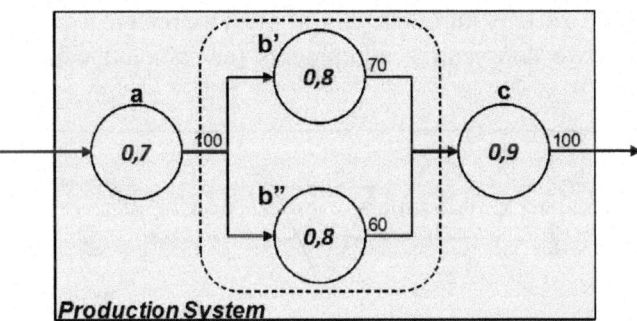

Figure 9: Availability of partially redundant equipments connected with not redundant equipments at modular capacity.

Table 5: State Analysis Configuration

b'	b''	Maximum Throughput	Probability of occurrence	[*100]	Expected Pieces Produced
UP	UP	100	0,8*0,8	0,64	64
UP	DOWN	70	0,8*(1-0,8)	0,16	11,2
DOWN	UP	60	(1-0,8)*0,8	0,16	9,6
	Expected number of Pieces Produced				84,8

Minor Stoppages and Speed Reduction

OEE theory includes in performance losses both the cycle time slowdown and minor stoppages. Also time losses of this category propagate, as stated before, throughout the whole production process.

A first type of performance losses propagation is due to the propagation of minor stoppages and reduced speed among machines in series system. From theoretical point of view, between two machines with the same cycle time [1] - and without buffer, minor stoppage and reduced speed propagate completely like as major stoppage. Obviously just a little buffer can mitigate the propagation.

Several models to study the role of buffers in avoiding the propagation of performance losses are available in *Buffer Design for Availability* literature [22]. The problem is of scientific relevance, since the lack of opportune buffer between the two stations can indeed affect dramatically the availability of the whole system. To briefly introduce this problem we refer to a production system composed of two consecutive equipments (or stations) with an interposed buffer (figure 10).

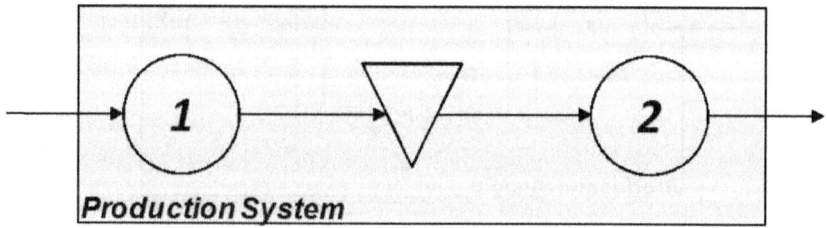

Figure 10: Station-Buffer-Station system. Adapted by [23].

Under the likely hypothesis that the ideal cycle times of the two stations are identical [23], the variability of speed that affect the stations is not necessarily of the same magnitude, due to its dependence on several factors. Furthermore Performance index is an average of the TtTt, therefore a same machine can sometimes perform at a reduced speed and sometimes an highest speed[2] - . The presence of this effect in two consecutive equipments can be mutually compensate or add up. Once again, within the propagation analysis for production system design, the role of buffer is dramatically important.

When buffer size is null the system is in series. Hence, as for availability, speed losses of each equipment affect the performance of the whole system:

$$P_{system} = \prod_{i=1}^{n} P_i$$

(18)

Therefore, for the two stations system we can posit:

$$P_{system} = \prod_{i=1}^{2} P_i$$

(19)

But when the buffer is properly designed, it doesn't allow the minor stoppages and speed losses to propagate from a station to another. We define this Buffer size as Bmax. When, in a production system of n stations, given any couple of consecutive station, the interposed buffer size is Bmax (calculated on the two specific couple of stations), then we have:

$$P_{system} = Min_{i=1}^{n}(P_i)$$

(20)

That for the considered 2 stations system is:

$$P_{system} = Min\ (P_1,\ P_2)$$

(21)

Hence, the extent of the propagation of performance losses depends on the buffer size (j) that is interposed between the two stations. Generally, a bigger buffer increases the performance of the system, since it increases the decoupling degree between two consecutive stations, up to j=Bmax is achieved (j =0,...,Bmax).

We can therefore introduce the parameter

$$Rel.P(j) = \frac{P(j)}{P(Bmax)}$$

(22)

Considering the model with two station, figure 11, we have that:

$$\textbf{When } j = 0, \ \ \textbf{Rel.P}(0) = \frac{P(0)}{P(Bmax)} = P(1)*P(2)/min(P(1); P(2));$$

(23)

When j = Bmax, Rel.P $\left(\mathbf{B}\ \mathbf{max}\right)$ = $\frac{P\,(Bmax)}{P(Bmax)}$ = 1; (24)

Figure 11 shows the trend of *Rel.P(j)* depending on the buffer size *(j)*, when the performance rate of each station is modeled with an exponential distribution [23] in a flow shop environment. The two curves represent the minimum and the maximum simulation results. All the others simulation results are included between these two curves. Maximum curve represents the configuration with the lowest difference in performance index between the two stations, the minimum the configuration with the highest difference.

By analyzing the figure 11 it is clear how an inopportune buffer size affect the performance of the line and how increase in buffer size allows to obtain improve in production line OEE. By the way, once achieved an opportune buffer size no improvement derives from a further increase in buffer. These considerations of Performance index trend are fundamental for an effective design of a production system.

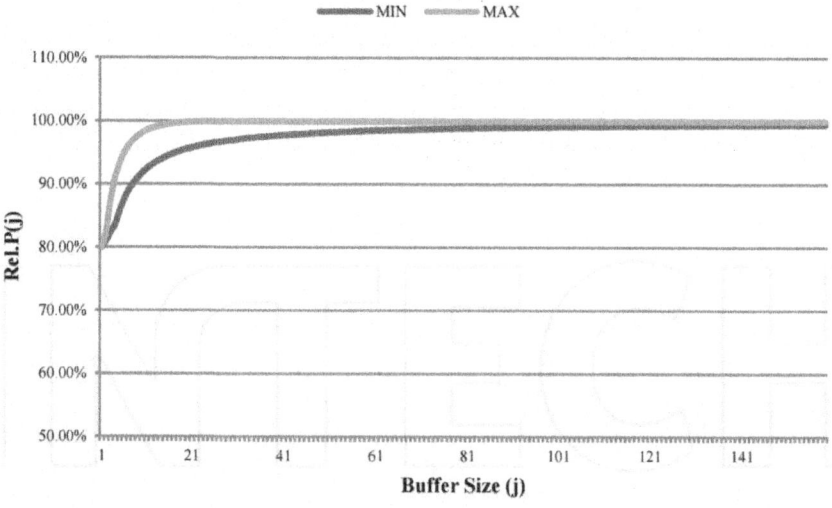

Figure 11: Rel OEE depending on buffer size in system affected by variability due to speed losses.

Quality Losses

In this paragraph we analyze how quality losses propagate in the system and if it is possible to assess the effect of quality control on OEE and OTE.

First of all we have to consider that quality rate for a station is usually calculated considering only the time spent for the manufacturing of products that have been rejected in the same station. This traditional approach focuses on stations that cause defects but doesn't allow to point out completely the effect of the machine defectiveness on the system. In order to do so, the total time wasted by a station due to quality losses should include even the time spent for manufacturing of good products that will be rejected for defectiveness caused by other stations. In this sense quality losses depends on where scraps are identified and rejected. For example, scraps in the last station should be considered loss of time for the upstream station to estimate the real impact of the loss on the system and to estimate the theoretical production capacity needed in the upstream station. In conclusion the authors propose to calculate quality rate for a station considering as quality loss all time spent to manufacture products that will not complete the whole process successfully.

From a theoretical point of view we could consider the following case for calculation of quality rate of a station that depends on types of rejection (scraps or rework) and on quality controls positioning. If we consider two stations with an assigned defectiveness S_j and each station reworks its scraps with a rework cycle time equal to theoretical cycle time, quality rate could be formulate as shown in case 1 infigure 12. Each station will have quality losses (time spent to rework products) due its own defectiveness. If we consider two stations with an assigned defectiveness S_j and a quality control station at downstream each station, quality rate could be formulate as shown in case 2 in figure 12. The station 1 that is the upstream station will have quality losses (time spent to work products that will be discarded) due to its own and station 2 defectiveness. If we consider two stations with an assigned defectiveness S_j and quality control station is only at the end of the line, quality rate quality rate could be formulate as shown in case 3 in figure 12. In this case both stations will have quality losses due to the propagation of defectiveness in the line. Case 2 and 3 point out that quality losses could be not simple to evaluate if we consider a long process both in design and management of system. In particular in the quality rate of station 1 we consider time lost for reject in the station 2.

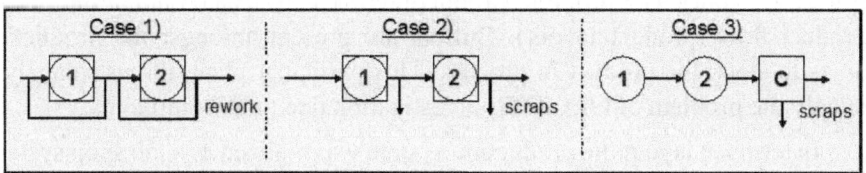

Figure 12: Different cases of quality rate calculation.

Finally, it is important to highlight the different role that the quality efficiency plays during the design phase and the production.

When the system is producing, Operations Manager focuses his attention on the causes of the delectability with the aim to reduce it. When it is to design the production system, Operations Manager focuses on the expected quality efficiency of each station, on the location of quality control, on the process (rework or scraps) to identify the correct number of equipments or station for each activity of the process.

In this sense, the analysis is vertical during the production phase, but it follows the whole process during the design (figure 13).

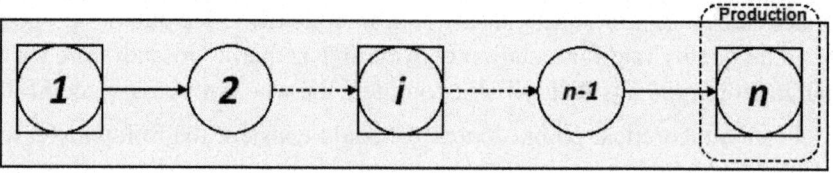

Figure 13: Two approaches for quality efficiency.

THE SIMULATION MODEL

To study losses propagation and to show how these dynamics affect OEE in a complex system [25] this chapter presents some examples taken from an OEE study of a real manufacturing system carried out by the authors through a process simulation analysis [19].

Simulation is run for each kind of time losses (Availability, Performance and Quality), to clearly show how each equipment ineffectiveness may compromise the performance of the whole system.

The simulation model is about a manufacturing plant for production of electrical cable. In particular we focuses on production of unipolar electrical cable that takes place by a flow-shop process. In the floor plant the production equipment is grouped in production areas arranged according to their functions (process layout). The different production areas are located along the line of product flow (product layout). Buffers are present amongst the production areas to stock the product in process. This particular plant allows to analyze deeply the problem of OEE-OTE investigation due to its complexity.

In terms of layout the production system was realized as a job shop system, although the flow of material from a station to another was continuous and typical of flow shop process. As stated in (§2) the reason lies on due to the huge size of the products that passes from a station to another. For this reason the buffer amid station, although present, couldn't contain huge amount of material.

The process implemented in the simulation model is shown in figure 14. Entities are unit quantity of cable that have different mass amongst stations. Parameters that are data input in the model are equipment speed, defectiveness, equipment failure rate and mean time to repair. Each parameters is described by a statistical distribution in order to simulate random condition. In particular equipment speed has been simulated with a triangular distribution in order to simulate performance losses due to speed reduction.

The model evaluates OTE and OEE for each station as usually measured in manufacturing plant. The model has been validated through a plan of tests and its results of OEE has been compared with results obtained from an analytic evaluation.

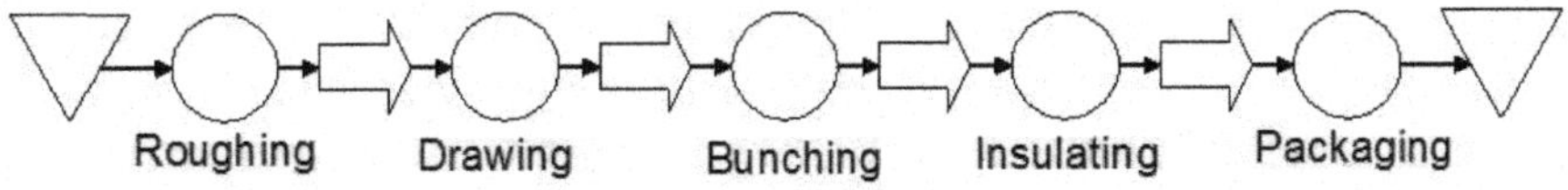

Roughing   Drawing   Bunching   Insulating   Packaging

Figure 14: ASME representation of manufacturing process.

Example of Availability Losses Propagation

In accordance with the proposed method (§ 3.5) we show how availability losses propagate in the system and to assess the effect of buffer capacity on OEE through the simulation. We focuses on the insulating and packaging working stations. Technical data about availability of equipment are: mean time between failure for insulating is 20000 sec while for packaging is 30000 sec; mean time between repair for insulating is 10000 sec while for packaging is 30000 sec. The cycle time of the working stations are the same equal to 2800 sec for coil. The quality rates are set to 1. Idling, minor stoppages and reduced speed are not considered and set to 0.

Considering equipment isolated from the system the OEE for the single machine is equal to its availability; in particular, relating to previous data, machines have an OEE equal to 0,67 and 0,5 respectively for insulating and packaging. The case points out how the losses due to major stoppage spread to other station in function of buffer capacity dimension.

A simulation has been run to study the effect of buffer capacity in this case. Capacity of buffer downstream of insulating has been changed from 0 to 30 coils for different simulations. The results of simulations are shown in figure 15a. The OEE for both machines is equal to 0,33 with no buffer capacity. This results is the composition of availability of insulating and packaging (0,67 x 0,5) as expected. The OEEs increase in function of buffer dimension that avoids the propagation of major stoppage and availability losses propagation.

Also the OTE is equal to 0,33 that is, according to formulation in (1) and as previously explained, equal to OEE of the last station but assessed in the system.

Insulating and packaging increase rapidly OEEs since a structural limits of buffer capacity of 15 coils; from this value OEEs of two stations converge on value of 0,5. The upstream insulating station, that has an availability greater than packaging, has to adapt itself to real cycle time of packaging that is the bottleneck station.

It's important to point out that in performance monitoring of manufacturing plant the propagation of the previous losses is often gathered as performance losses (reduced speed or minor stoppage) in absence of specific data collection relating to major stoppage due to absence of material flow. So, if we consider also all other efficiency looses ignored in this sample, we can understand how much could be difficult to identify the real impact of this kind of efficiency losses monitoring the real system. Moreover simulation supports in system design in order to dimension buffer capacity (e.g. in this case structural limit for OEE is reached for 16 coils). Moreover through simulation it is possible to point out that the positive effect of buffer is reduced with an higher cycle time of machine as shown infigure 15b.

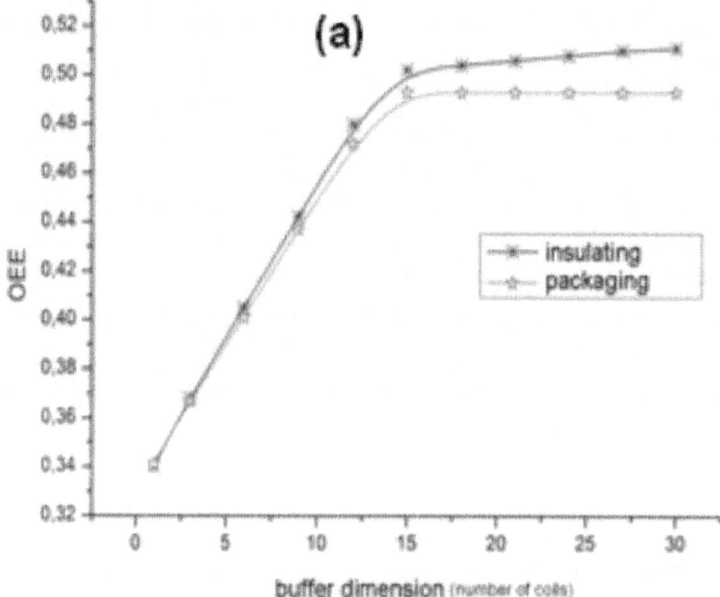

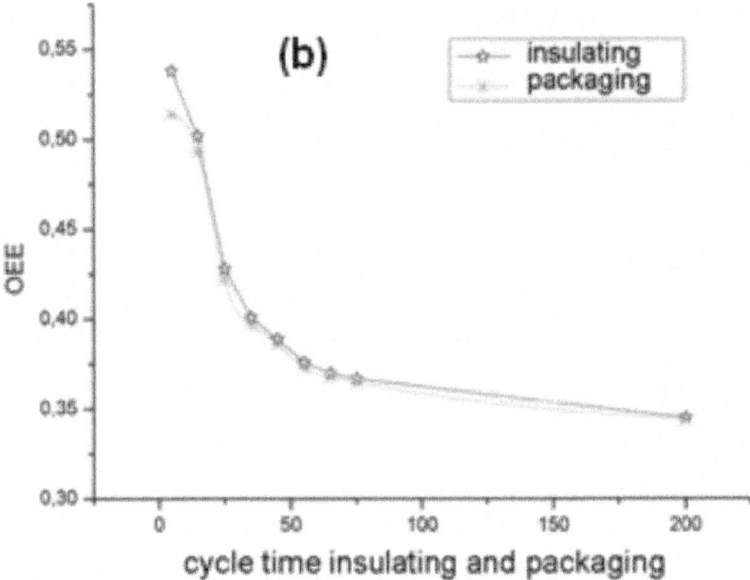

Figure 15: OEE in function of buffer dimension (a) and cycle time (b).

Minor Stoppages and Speed Reduction

We run the simulation also for the case study (§ 4). The simulation shown how two stations, with the same theoretical cycle time (200 sec/coil) affected by a triangular distribution with a performance rate of 52% as single machine, have: 48% of performance rate with a capacity buffer of 1 coil and 50% of performance rate with a capacity buffer of 2 coils. But if we consider two stations with the same theoretic cycle time but affects by different triangular distributions so that theoretic performance rates differ, simulation shows how the performance rates of two stations converge towards the lowest one as expected (19), (20).

Through the same simulation model we considered also the second type of performance losses propagation, due to the propagation of reduced speed caused by unbalanced line. Figure 16 shown the effect of unbalanced cycle time of stations relating to insulating and packaging. The station have the same P as single machine equal to 67% but different theoretical cycle time. In particular insulating, the upstream station, is faster than packaging. Availability and quality rate of stations is set to 1. The buffer capacity is set to 1 coil. A simulation has been run to study the effect of unbalancing station. Theoretical cycle time of insulating has been changed since theoretical cycle time of packaging that is fixed in mean. The simulation points out that insulating has to adapt itself to cycle time of packaging that is the bottleneck station. This results in the model as a lower value for performance rate of insulating station. The same happens often in real systems where the result is influenced by all the efficiency losses at the same time. The effect disappears gradually with a better balancing of two stations as in figure 16.

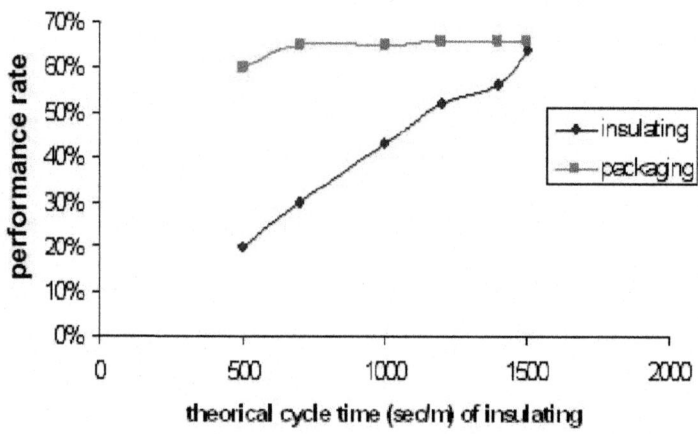

Figure 16: Performance rate of insulating and packaging in function of insulating cycle time.

Quality Losses

In relation to the model, this sample focuses on the drawing and bunching working stations that have defectiveness set to 5%, the same cycle times and no other efficiency losses. The quality control has been changed simulating case 2 and 3. The results of simulation for the two cases are shown in table 6 in which the proposal method has compared with the traditional one. The proposal method allowed to identify the correct efficiency, for example to dimension the drawing station, because it considers time wasted to manufacture products rejected in bunching station. The difference between values of Q_2 and OTE is explained by the value of $P_2=0,95$ that is due to the propagation of quality losses for the upstream station in performance losses for the downstream station. Moreover about positioning of quality control the case 2 has to be prefer because the simulation shows a positive effect on the OTE if the bunching station is the system bottleneck (as it happens in the real system).

Table 6: Comparison of quality rate calculation and evaluation of impact of quality control positioning on quality rates and on OTE

	Proposal method			Traditional method		
	Q1	Q2	OTE	Q1	Q2	OTE
Case 2)	$0,9^52$	0,95	$0,9^52$	0,95	0,95	$0,9^52$
Case 3)	$0,9^52$	$0,9^52$	$0,9^52$	--	$0,9^52$	$0,9^52$

CONCLUSIONS

The evaluation of Overall Equipment Effectiveness (OEE) and Overall Throughput Effectiveness (OTE) can be critical for the correct estimation of workstations number needed to realize the desired throughput (production system design), as also for the analysis and the continuous improvement of the system performance (during the system management).

The use of OEE as performance improvement tool has been widely described in the literature. But it has been less approached in system design for a correct evaluation of the system efficiency (OTE), in order to study losses propagation, overlapping of efficiency losses and effective actions for losses reduction.

In this chapter, starting by the available literature on time losses, we identified a simplified set of relevant time-losses that need to be considered during the design phase. Then, through the simulation, we shown how OEE of single machine and the value of OTE of the whole system are interconnected and mutually influencing each other, due to the propagation of availability, performance and quality losses throughout the system.

For each category of time losses we described the effects of efficiency losses propagation from a station to the system, for a correct estimation and analysis of OEE and OTE during manufacturing system design. We also shown how to avoid losses propagation through adequate technical solutions which can be defined during system design as the buffer sizing, the equipment configuration and the positioning of control stations.

The simulation model shown in this chapter was based on a real production system and it used real data to study the losses propagation in a manufacturing plant for production of electrical cable. The validation of the model ensures the meaningful of the approach and of the identified set of possible solutions and hints.

By analyzing and each time losses we also shown how the choices taken during the design of the production system to increase the OTE (e.g. buffer size, maintenance configuration, etc.) affect the successive management of the operations.

ACKNOWLEDGEMENTS

The realization of this chapter would not have been possible without the support of a person whose cooperated with the chair of Operations Management of University of Rome "Tor Vergata" in the last years, producing valuable research. The authors wish to express their gratitude to Dr. Bruna Di Silvio without whose knowledge, diligence and assistance this work would not have been successful.

Notes

[1] - As shown in par. 3.1. When two consecutive stations present different cycle times, the faster station works with the same cycle time of slower station, with consequence on equipment OEE, even if any time losses is occurred. On the other hand, when two consecutive stations are balanced (same cycle time) if any time loss is occurring the two stations OEE will be 100%. Ideally, the higher value of performance rate can be reached when the two stations are balanced.

[2] - This time losses are typically caused by yield reduction (the actual process yield is lower than the design yield). This effect is more likely to be considered in production process where the equipment saturation level affect its yield, like furnaces, chemical reactor, etc.

REFERENCES

1.   H. H. S., «Manufacturing productivity improvement using effectivenes metrics and simulation analysis,» 2002.

2.   B. I., «Effective measurement and successful elements of company productivity: the basis of competitiveness and world prosperity,» *International Journal of Production Economics,* vol. 52, pp. 203-213, 1997.

3.   P. D Jeong, K.Y., «Operational efficiency and effectiveness measurement,» *International Journal of Operations and Production Management, 21*n. 1404-1416, 2001

4.   N. S., Introduction to TPM- Total Productive Maintenance, Productivity Press, 1988.

5.   S. R.J., World Class Manufacturing. The lesson of simplicity Applied, The Free Press, 1987.

6.   J. D Womack, J.P., Lean Thinking, Simon & Schuster, 1996

7.   N. A. V Dixon, J.R., The new performance challenge. Measuring operations for world-class competition, Dow Jones Irwin, 1990

8.   S. D., «Can CIM improve overall factory effetivenes,» in *Pan Pacific Microelectronic Symposium*, Kauai, HI, 1999.

9.   W. D Waters, D.J., Operations Management, Kogan Page Publishers, 1999

10. A. N. J. F Chase, R.B., Operations Management, McGraw-Hill, 2008

11. A. V. A., Semiconductor Manufacturing Productivity- Overall Equipment Effectiveness (OEE) guidebook, SEMATECH, 1995.

12. D. R. A. J Rooda, J.E., «Equipment effectiveness: OEE revisited,» *IEEE Transactions on Semiconductor Manufacturing, 18* n. 1, 189 196 2005

13. G. L. R. B Gamberini, R., «Alternative approaches for OEE evaluation: some guidelines directing the choice,» in *XVII Summer School Francesco Turco*, Venice, 2012

14. T. F Grando, A., «Modelling Plant Capacity and Productivity,» *Production Planning and Control, 16* n. 3, 209 322 2005

15. C. V Spada, C., «The Impact of Cultural Issues and Interpersonal Behavior on Sustainable Excellence and Competitiveness: An Analysis of the Italian Context,» *Contributions to Management Science, 95 113* 2008

16. G. R Badiger, A. , «A Proposal, evaluation of OEE and impact of six big losses on equipment earning capacity,» *International Journal Process Management & Benchmarking, 235 247* 2008

17. A. N. C. B Jacobs, R. F Operations, and supply chain management, McGraw-Hill, A cura di, 2010

18. R Oechsner, From overall equipment efficiency(OEE) to overall Fab effectiveness (OFE),» *Materials Science in Semiconductor Processing, 5* 333 339 2003

19. D. S. B. C. V Introna, V Flow-shop, process oee calculation and improvement using simulation analysis,» in *MITIP*, Florence, 2007

20. R. MA, «Factory Level Metrics: Basis for Productivity Improvement,» in *Proceedings of the International Conference on Modeling and Analysis of Semiconductor*, Tempe, Arizona, USA, 2002.

21. P. R Scott, D., «Can overall factory effetiveness prolong Moore's Law?,» *Solid State Technology, 41 75 82* 1998

22. B. D., «Buffer size design linked to reliability performance: A simulative study,» *Computers & Industrial Engineering,* vol. 56, p. 1633-1641,

2009.

23. G. A. V Introna, V., «Increasing Availability of Production Flow lines through Optimal Buffer Sizing: a Simulative Study,» in *The 23rd European Modeling & Simulation Symposium (Simulation in Industry)*, Rome, 2011

24. P. D. T. O Connor, Practical Reliability Engineering (Fourth Ed.), New York: John Wiley & Sons, 2002

25. J. O Kane, Simulating production performance: cross case analysis and policy implications,» *Industrial Management & Data Systems, 104* n. 4, 309 321 2004

26. B. A Gondhinathan, R. , «A Proposal, evaluation of OEE and impact of six big losses on equipment earning capacity,» *International Journal Process Management & Benchmarking, 2* n. 3, 235 247 2008

Chapter 10

MATERIALS HANDLING IN FLEXIBLE MANUFACTURING SYSTEMS

Dr. Tauseef Aized

Professor, Department of Mechanical, Mechatrnics and Manufacturing Engineering, KSK Campus, University of Engineering and Technology, Lahore, Pakistan

INTRODUCTION

Material handling can be defined as an integrated system involving such activities as moving, handling, storing and controlling of materials by means of gravity, manual effort or power activated machinery. Moving materials utilize time and space. Any movement of materials requires that the size, shape, weight and condition of the material, as well as the path and frequency of the move be analyzed. Storing materials provide a buffer between operations. It facilitates the efficient use of people and machines and provides an efficient organization of materials. The considerations for material system design include the size, weight, condition and stack ability of materials; the required throughput; and building constraints such as floor loading, floor condition, column spacing etc. The protection of materials include both packaging and protecting against damage and theft of material as well as the use of safeguards on the information system to include protection against the material being mishandled, misplaced, misappropriated and processed in a wrong sequence. Controlling material includes both physical control as well as status of material control. Physical control is the orientation of sequence and space between material movements. Status control is the real time awareness of the location, amount, destination, origin, ownership and schedule of material. Maintaining the correct degree of control is a challenge because the right amount of control depends upon the culture of the organization and the people who manage and perform material handling functions.

Material handling is an important area of concern in flexible manufacturing systems because more than 80 % of time that material spends on a shop floor is spent either in waiting or in transportation, although both these activities are non-value added activities. Efficient material handling is needed for less congestion, timely delivery and reduced idle time of machines due to non-availability or accumulation of materials at workstations. Safe handling of materials is important in a plant as it reduces wastage, breakage, loss and scrapes etc.

PRINCIPLES OF MATERIAL HANDLINGS

The material handling principles provide fundamentals of material handling practices and provide guidance to material handling system designers. The following is a brief description of material handling principles.

Planning Principle

All material handling should be the result of a deliberate plan where the needs, performance objectives and functional specification of the proposed methods are completely defined at the outset. In its simplest form a material handing plan defines the material (what) and the moves (when and where); together they define the method (how and who).

Standardization Principle

Standardize handling methods and equipments wherever possible. Material handling methods, equipment, controls and software should be standardized within the limits of achieving overall performance objectives and without sacrificing needed flexibility, modularity and throughout anticipation of changing future requirements.

Ergonomic Principle

Human capabilities and limitations must be recognized and respected in the design of material handling tasks and equipment to ensure safe and effective operations. Equipments should be selected that eliminates repetitive and strenuous manual labor and which effectively interacts with human operators and users.

Flexibility Principle

Use methods and equipments that can perform a variety of tasks under varying operating conditions.

Simplification

Simplify material handling by eliminating, reducing or combining unnecessary movements and equipments.

Gravity

Utilize gravity to move material wherever possible.

Layout

Prepare an operation sequence and equipment layout for all viable system solutions and then select the best possible configuration.

Cost

Compare the economic justification of alternate solutions with equipment and methods on the basis of economic effectiveness as measured by expenses per unit handled.

Maintenance

Prepare a plan for preventive maintenance and scheduled repairs on all material handling equipments.

Unit Load Principle

A unit load is one that can be stored or moved as a single entity at one time, such as a pallet, container or tote, regardless of the number of individual items that make up the load. Unit loads shall be appropriately sized and configured in a way which achieves the material flow and inventory objectives at each stage in the supply chain.

Space Utilization Principle

Effective and efficient use must be made of all available space. In work areas, cluttered and unorganized spaces and blocked aisles should be eliminated. When transporting loads within a facility, the use of overhead space should be considered as an option.

System Principle

Material movement and storage activities should be fully integrated to form a coordinated, operational system which spans receiving, inspection, storage, production, assembly, packaging, unitizing, order selection, shipping, transportation and the handling of returns. Systems integration should encompass the entire supply chain including reverse logistics. It should include suppliers, manufacturers, distributors and customers.

Automation Principle

Material handling operations should be mechanized and/or automated where feasible to improve operational efficiency, increase responsiveness, and improve consistency and predictability.

Environmental Principle

Environmental impact and energy consumption should be considered as criteria when designing or selecting alternative equipment and material handling systems.

Life Cycle Cost Principle

A thorough economic analysis should account for the entire life cycle of all material handling equipment and resulting systems. Life cycle costs include capital investment, installation, setup and equipment programming, training, system testing and acceptance, operating (labor, utilities, etc.), maintenance and repair, reuse value, and ultimate disposal.

MATERIAL TRANSPORT EQUIPMENT

International Materials Management Society has classified equipment as (1) conveyor, (2) cranes, elevators, and hoists, (3) positioning, weighing, and control equipment, (4) industrial vehicles, (5) motor vehicles, (6) railroad cars, (7) marine carriers, (8) aircraft, and (9) containers and supports. The following provides the details of material transport equipments.

Conveyor Systems

A Conveyor is used when a material is moved very frequently between specific points and the path between points is fixed. Conveyors combined with modern identification and recognition systems like bar code technologies have played a significant role in the transportation and sorting of a large variety of products in modern warehouses. Some of the common types of conveyors are:

- Roller conveyor
- Skate- wheel conveyor
- Belt conveyor

- In- floor towline conveyor
- Overhead trolley conveyor
- Cart-on-track conveyor

Roller Conveyor

In roller conveyors, the pathway consists of a series of rollers that are perpendicular to the direction of travel. Loads must possess a flat bottom to span several rollers which can be either powered or non-powered. Powered rollers rotate to drive the loads forward in roller conveyor. The following figure shows a roller conveyor.

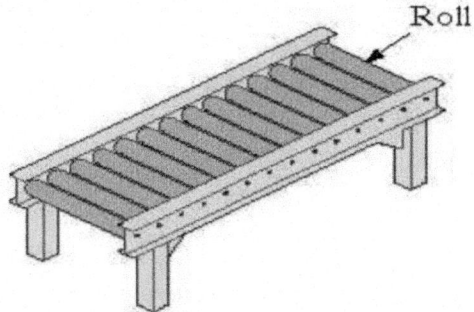

Figure 1: Roller conveyor.

Skate-Wheel Conveyor

Skate-wheel conveyors are similar in operation to roller conveyor but use skate wheels instead of rollers and are generally lighter weight and non-powered. Sometimes, these are built as portable units that can be used for loading and unloading truck trailers in shipping and receiving. Figure 2 shows a skate-wheel roller.

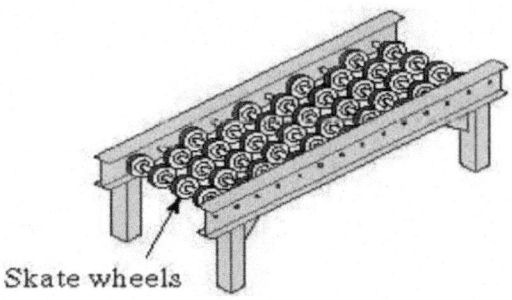

Figure 2: Skate-wheel conveyor.

Belt Conveyor

A belt conveyor is a continuous loop with forward path to move loads in which the belt is made of reinforced elastomeric support slider or rollers used to support forward loop. There are two common forms:

- Flat belt (shown)
- V-shaped for bulk materials

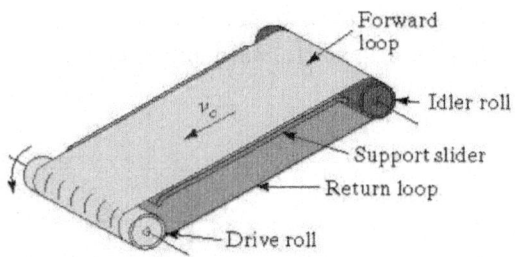

Figure 3: Belt conveyor.

In-Floor Tow-Line Conveyor

These are four-wheel carts powered by moving chains or cables in trenches in the floor. Carts use steel pins (or grippers) to project below floor level and engage the chain (or pulley) for towing. This allows carts to be disengaged from towline for loading and unloading purpose as is shown in Figure 4.

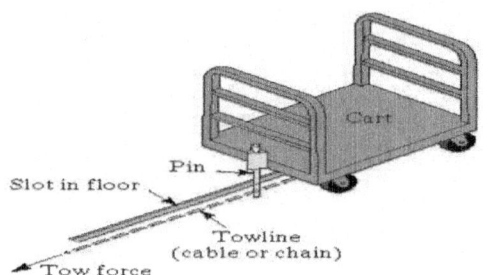

Figure 4: In-floor two-line conveyor.

Overhead Trolley Conveyor

A trolley is a wheeled carriage running on an overhead track from which loads can be suspended. Trolleys are connected and moved by a chain or cable that forms a complete loop and are often used to move parts and assemblies between major production areas. Figure 5 shows an overhead trolley conveyor.

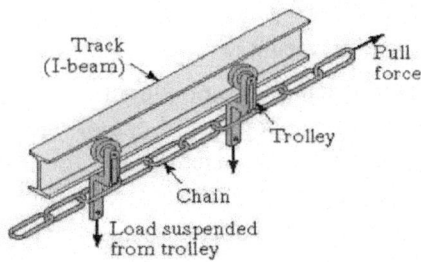

Figure 5: Over-head trolley conveyor.

Cart-On-Track Conveyor

Carts ride on a track above floor level and are driven by a spinning tube. The forward motion of cart is controlled by a drive wheel whose angle can be changed from zero (idle) to 45 degrees (forward). It is shown in the following figure.

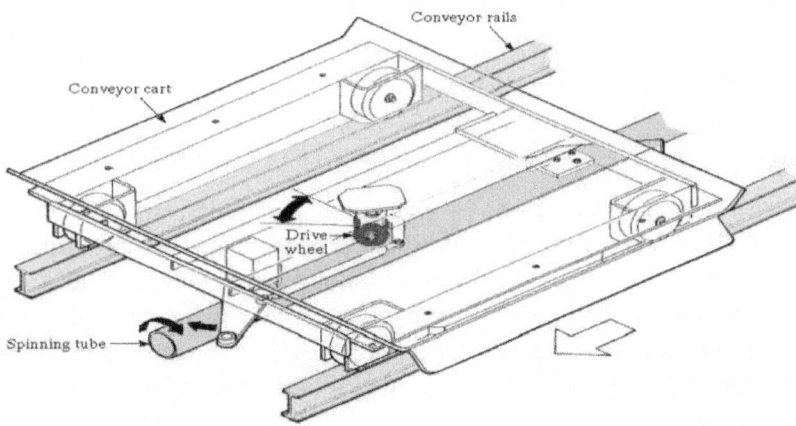

Figure 6: Cart-on-track coveyor.

Cranes and Hoists

Cranes are normally used for transferring materials with some considerable size and weight and for intermittent flow of material. In general, loads handled by cranes are more varied with respect to their shape and weight than those handled by a conveyor. Hoists are frequently attached to cranes for vertical translation that is, lifting and lowering of loads. They can be operated manually, electrically, or pneumatically. Cranes usually include hoists so that the crane-and-hoist combination provides

- Horizontal transport
- Vertical lifting and lowering

This class of material handling equipments can typically lift & move a material up to 100 tons. A hoist consists of one or more fixed pulley & one or more rotatable pulley & a hook to attach load with it. The number of pulleys in hoist determines its mechanical advantage which is the ratio of load lifted & deriving force. Hoist with mechanical advantage of four are shown below:

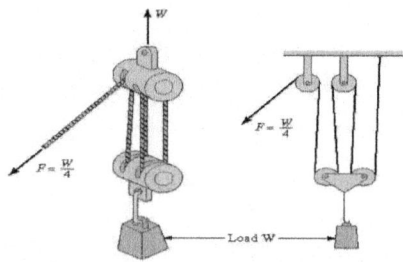

Figure 7: (a) Sketch of the hoist (b) diagram to illustrate mechanical advantage.

There are different types of cranes that are used in industrial applications. Some of these are discussed below.

Bridge Crane

A bridge crane consist of one or two horizontal girder or beam suspended between fixed rail on either end which are connected to the structure of building. The hoist trolley can be moved along the length of bridge & bridge can be moved the length of rail in building. These two capabilities provide motion along X-axis & Y-axis whereas hoist can provide motion in the z-axis. Their application includes heavy machinery fabrication. They have ability to carry load up to 100 tons.

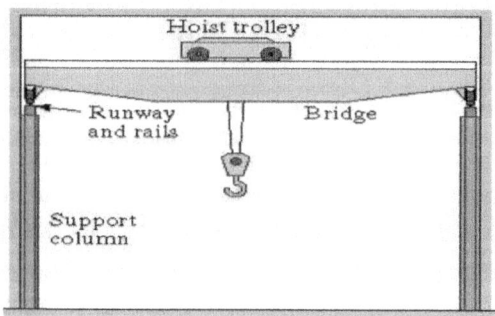

Figure 8: Bridge crane.

Half-Gantry Crane

Half gantry crane is distinguished from bridge crane by the presence of one or two vertical supporting elements which support horizontal girder. Gantry cranes may be half or double. Half gantry has one supporting vertical element whereas double gantry crane has two vertical supporting legs.

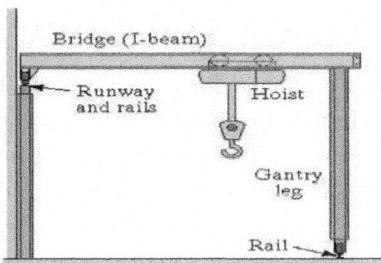

Figure 9: Half gantry crane.

Jib Crane

Jib cranes consist of a rotating arm with a hoist that runs along its length. The arm usually revolves on an axis which can be a fixed, ground-mounted post, or can be a wall or ceilingmounted pin.

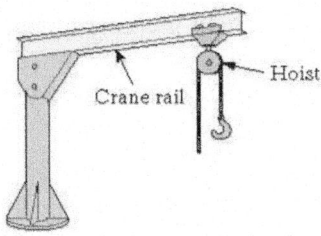

Figure 10: Jib Crane.

Wall-bracket mounted jib cranes are usually the least expensive jib cranes, but they require the most headroom and exert more force on their mounting wall. Cantilever jib cranes place the arm at the top, allowing for maximum lift when used in situations with limited headroom. They also exert less force on the wall on which they're mounted. Tie rod jib cranes make use of a tie rod between the arm and the mounting area. More inexpensive jib cranes feature manually operated chain hoists, while sophisticated cranes use an electric chain hoist. Jib cranes are used when the desired lifting area resides within a semi- circular arc.

Stacker Crane

It is similar to a bridge crane. The major difference is that, instead of using a hoist, the stacker crane uses a mast with forks or a platform to handle unit loads. Stacker cranes are generally used for storing and retrieving unit loads in storage racks, especially in high-rise applications.

AUTOMATED RETRIEVAL AND STORAGE EQUIPMENTS

Storage equipments can be in the form of racks, shelves, bins and drawers. Among these, storage rack is probably the most common form of storage equipment. There are numerous variants and configurations of storage racks, which include single-deep, double-deep rack, cantilever rack etc. and configurations that are designed to facilitate specific storage and retrieval operations drive-through, flow-through etc. More sophisticated retrieval and storage system combine the use of storage equipment, storing and retrieval machines and control that are manifested in a modern automated storage/retrieval system.

Automated Guided Vehicles

An Automated Guided Vehicle System (AGVS) is a material handling system that uses independently operated, self-propelled vehicles guided along defined pathways in the facility floor. It is an automated material handling system which moves along predefined and preprogrammed path along an aisle from one station to another. The main parts of an AGV include structure, drive system, steering mechanism, power source (battery) and onboard computer for control.

Types of AGV

The following are common types of AGVs.

Driverless Automated Guided Train

These are the first type of AGVS to be introduced around 1954.Its typical application is moving heavy payloads over long distances in warehouses and factories without intermediate stops along the route

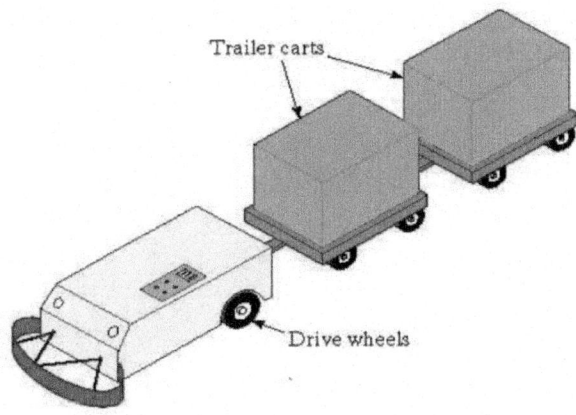

Figure 11: Driverless automated guided vehicle.

AGV Pallet Truck

These are used to move palletized loads along predetermined routes. Vehicle is backed into loaded pallet by worker; pallet is then elevated from floor. Worker drives pallet truck to AGV guide path and programs destination.

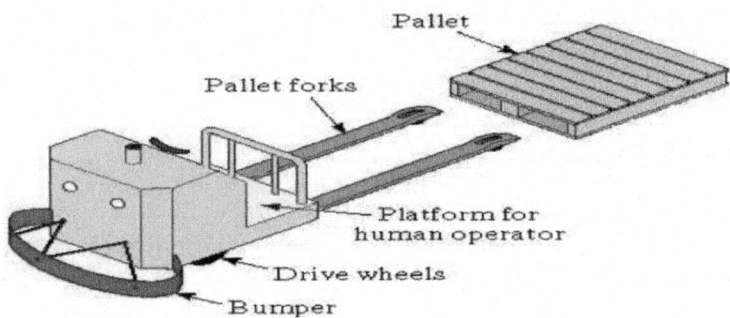

Figure 12: AGV pallet truck.

Unit Load Carrier

These are used to move unit loads from station to station and are often equipped for automatic loading/unloading of pallets using roller conveyors, moving belts, or mechanized lift platforms.

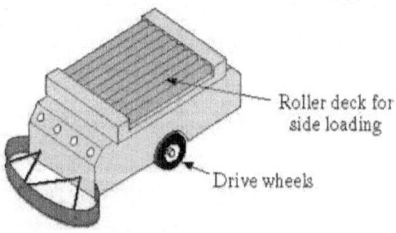

Figure 13: Unit load carrier.

Light Load AGV

It can be applied for smaller loads. These are typically used in electronics assembly and office environments as mail and snack carriers.

Assembly AGV

These are used as assembly platforms, for example car chassis, engines etc., by carrying products and transport them through assembly stations.

Forklift AGV

It has the ability to pick up and drop off palletized loads both at floor level and on stands. Generally, these fork lift AGVs have sensors on forks for pallet interfacing.

Rail-Guided Vehicles

These are self-propelled vehicles that ride on a fixed-rail system. These vehicles operate independently and are driven by electric motors that pick up power from an electrified rail. Fixed rail system may be:

- Overhead monorail - suspended overhead from the ceiling
- On-floor - parallel fixed rails, tracks generally protrude up from the floor

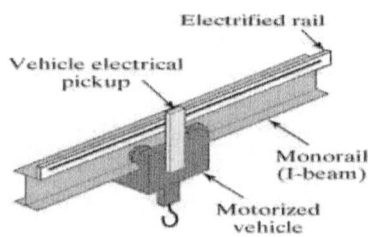

Figure 14: Rail guided vehicle.

AGVS System Management

AGVS is a complex system and a number of parameters need to be considered which include:

Guide-path layout

Number of AGVs required

Operational and transportation control

Guide-Path Layout

The guide-path layout defines the possible vehicle movement path. Links and nodes that represent the action points such as pick-up and drop-off points, maintenance areas and intersections represent the path. The guide-path can be divided into four types:

1. Unidirectional single lane guide-path
2. Bi-directional single lane guide-path
3. Multiple lanes
4. Mixed guide-path.

Generally bidirectional single lane is considered the most cost effective and widely used layout.

Number of AGVs required

It is important to estimate the optimum number of AGVs required for a system as too many AGVs will congest the traffic while too few means larger idle time for workstations in a system. Generally, the number of AGVs required is the sum of the total loaded and empty travel time and waiting time of the AGVs divided by the time an AGV is available.

Operational and Transportation Control

The operation and transportation consists of vehicle dispatching, vehicle routing and traffic control issues. Once a demand arises for an AGV, a choice needs to be made regarding the vehicle to be dispatched among the pool of vehicles available. In an event when several workstations need servicing, a choice is to be made as to which workstation is to be serviced. The selection criteria can be applied for assigning the vehicles or workstations based on one or a combination of the following:

- A random vehicle
- Longest idle vehicle

- Nearest vehicle
- Farthest vehicle
- Least utilized vehicle
- Random workstation
- Nearest workstation
- Farthest workstation
- Maximum queue size
- Minimum remaining queue size
- First come fist served
- Unit load arrival time, due time or priority.

In order to dispatch an AGV to any workstation, it is necessary to find the shortest feasible path from the existing position. While selecting the shortest path it is necessary to consider only those paths which are free and not occupied by vehicles. It may also be necessary to consider the future positions of the vehicles in the route in addition to their current occupied positions. In identifying the traffic control systems for AGVs movement, the approaches that can be used are forward sensing control, zone sensing control and combinatorial control. In forward sensing control, an AGV is equipped with obstruction detecting sensors that can identify another AGV in front of it and slow down or stop. This helps in improving the AGV utilization due to closer allowable distance between vehicles. However, this approach may not be able to detect the obstacles at intersections and around corners. This is generally useful for long and straight path which is divided into zones. Once an AGV enters a zone, it becomes unavailable for other AGVs which may introduce system inefficiency. The main advantages derived from the use of AGVs in manufacturing environment are: Dispatching, tracking and monitoring under real time control which help in planned delivery.

- Better resource utilization as AGVs can be economically justified.
- Increased control over material flow and movement
- Reduced product damage and routing flexibility
- Increased throughput because of dependable on-time delivery.

INDUSTRIAL ROBOTS

Industrial robots are very useful material handling devices in an automated environment. An industrial robot is a reprogrammable multifunctional manipulator designed to move materials, parts, tools, or other devices by means of variable programmed motions and to perform a variety of other tasks. It is also defined as a machine formed by a mechanism including several degrees of freedom often having the appearance of one or several arms ending in a wrist capable of holding a job, tool and inspection device. It is automatically controlled, reprogrammable, multipurpose manipulative machine with several reprogrammable axes which is either fixed in place or mobile for use in industrial automation applications.

Robot Components

The following are basic components of an industrial robot.

Manipulator

It is a mechanical unit that provides motions similar to those of human arm and hand. The end of wrist can reach a point in space having a specific set of coordinates in specific orientation.

End Effector

It is attached with the end of wrist in a robot. It is a special purpose tooling which enables the robot to perform a particular job. Depending on the type of work, end effector may be equipped with any of the following:

- Grippers, hooks, vacuum cups, and adhesive fingers for material handling
- Spray guns for painting
- Attachments for different kinds of welding processes.

Control System

It is a brain of a robot which gives commands for the movements of the robot. It stores the data to initiate and terminate movements of the manipulator. It interfaces with the computers and other equipments such as manufacturing cells or assembly operations.

Power Supply

It supplies the power to the controller and manipulator. Each motion of manipulator is controlled and regulated by actuators that use an electrical, pneumatic or hydraulic power.

Robot Types

Robots are generally classified as Cartesian or rectilinear, cylindrical, polar or spherical jointed arms. They are also classified, from material handling point of view, as under:

Pick and Place Robot

It is also called fixed sequence robot and is programmed for a specific operation. Its movements are from point to point and cycle is repeated. These robots are simple and inexpensive and are used to pick and place materials.

Playback Robot

This robot learns the work and motions from operator who leads the playback robot and its end effector through the desired path. The robot memorizes and records the path and sequence of motions and can repeat them continuously without any further action or guidance by the operator.

Numerically Controlled Robot

It is a programmable type of robot and works same as the numerical control machines. The robot is servo controlled by digital data and its sequence of movements can be changed with relative ease.

Intelligent Robot

It is capable of performing some of the functions and tasks carried out by humans and is equipped with a variety of sensors with usual and tactile capabilities. It can perform tasks such as moving among a variety of machines on a shop floor avoiding collisions. It can recognize, select and properly grip the correct work piece.

Robot Applications in Material Handling

The major applications in material handling include:

- Industrial robots are used to load/ unload materials during operations.
- These are used to transfer the material from one conveyor to another.
- These are used in palletizing and de-palletizing in such a way that parts/ materials are taken from conveyor and are loaded on to a pallet in a desired pattern and sequence and vice-versa.
- These are very effective in automated assembly where repetitive work is required.

- Intelligent robots can be used to automatically pick the right work piece without interference of operator and hence improves quality and pace of work.

REFERENCES

1. M.P. Groover. "Automation, Production systems and computer integrated manufacturing" Second edition. Pearson-Prentice Hall, 2008.

2. K. Sareen and C. Grewal."CAD/CAM: Theory and concepts" S. Chand & Co. 2009.

3. C. R. Alavala. " CAD/CAM: Concepts and applications" Prentice-Hall, 2008.

4. P. N. Rao. " CAD/CAM: Principles and applications" McGraw-Hill, 2004.

5. C. R. Asfahl. "Robots and manufacturing automation" Second edition, John-Wiley and sons.1992.

6. M. P. Groover and E. W. Zimmers. Jr. " CAD/CAM: Computer added design and manufacturing" Pearson-Prentice Hall, 2009.

7. G. Chryssolouris, "Manufacturing systems: Theory and Practice" Springer-Verlag,1992.

Chapter 11

MULTIDIMENSIONAL OF MANUFACTURING TECHNOLOGY, ORGANIZATIONAL CHARACTERISTICS, AND PERFORMANCE

Tritos Laosirihongthong

University Campus STeP Ri Slavka Krautzeka 83/A 51000 Rijeka, Croatia

INTRODUCTION

Over the last ten years manufacturing technology use has been studied in several countries and a stream of findings has been coming in. The purpose of this study is to investigate manufacturing technology use in the Thai automotive industry, and to (1) examine findings concerning certain manufacturing technology dimensions, (2) investigate the relationships among manufacturing technology use, organizational characteristics (i.e. size, ownership and unionization), and performance, and (4) use the findings to shape a concept of multidimensional view of manufacturing technology. In the past, many studies have used data from the US, Australia, and other developed countries (Boyer et, al., 1997; Sohal, 1999; Dean et, al, 2000: Park, 2000). The findings from this study using data of the Thai automotive industry are a useful contribution to international applicability of manufacturing technology.

This chapter is organized into five sections. The next section summarizes the literature and theoretical background. Research methodology and data analysis incorporating sample selection, questionnaire design, and reliability and validity of measurement instruments is described in Section 3. Research findings and conclusion is presented in Section 4 and 5 respectively.

LITERATURE REVIEW AND THEORETICAL BACKGROUND

Technology Dimensions

Certain classes of manufacturing technology are appropriate for particular competitive manufacturing strategy. For example, computer numerical control (CNC), computer-aided design (CAD), computer-aided manufacturing (CAM) or computer-aided engineering (CAE) are appropriate for a strategy seeking flexibility. Manufacturing technologies have been grouped and classified in several different ways, some based on the level of integration, or the nature of the technology. (Rosenthal, 1984; Warner, 1987; Adler, 1988; Paul and Suresh, 1991; Small and Chen, 1997).

Swamidass and Kotha (1998), in an empirical study, found that nineteen technologies used in manufacturing could be classified into four groups based on the volume and variety considerations of the production process. Their empirical results indicate that manufacturing technology could be classified into four groups:

- Information exchange and planning technology
- Product design technology
- High-volume automation technology and
- Low-volume flexible automation technology.

A notable conclusion of their study being that High-volume automation technology could be used to serve the low variety and high volume production strategy, while Product design technology and Low-volume flexible automation technology could be used to serve the high variety and low volume production strategy. The implication is that technology dimensions have far reaching consequences for the manner in which companies use them. This study decides to use the empirically-established dimensions of manufacturing technology reported by some previous studies, as described in section 3, to guide this study.

Manufacturing Technology Use and Organizational Characteristics

A number of previous studies have indicated that organizational characteristics (i.g., firm size, ownership, year in operation, sales volume, and labor union membership) have an influence on the adoption and implementation of manufacturing technology (Ettlie, 1984; Chen et al, 1996; Millen and Sohal, 1998; Schroder and Sohal, 1999; Swamidass and Winch, 2002). Summary of these findings are explained as follow:

Size

Manufacturing and operations management researchers have found that large companies show a higher degree of manufacturing technology implementation than small and medium companies (Paul and Suresh, 1991; Mansfield, 1993; Sohal, 1999; Swamidass and Kotha 1998). This is attributed in the literature to the fact that large companies have superior technological know-how because of their access to more human, financial and information resources compared to small to medium companies. Researchers have come to agree that size is an important variable when it comes to manufacturing technology use. For example, Small and Yasin (1997) recommend that future research in management of manufacturing technology should adopt a contingency approach to find out how organizational variables such as firm size, industry structure, and planning approach influence the relationship between adoption of manufacturing technology and overall plant performance.

The Nationality of Plant Ownership

Although a number of studies to investigate the relationship between organizational variables and technology use have been conducted in developed countries, such studies are not common in developing countries. Peter et al, (1999) state that the nationality of ownership of companies reflects the differences in management practice in manufacturing technology implementation due to differences in national culture. Sohal (1994) reports a number of significant differences in manufacturing technology use (e.g. computer hardware, computer software, plant and equipment) and management effort (e.g. source of manufacturing technology information, financial appraisal techniques, training, and benefits) between Australia and the United Kingdom. Lefley and Sarkis (1997) studied appraisal/assessment of manufacturing technology capital projects in the USA and UK and found different degrees of success in manufacturing technology implementation. Kotha and Swamidass (1998) report a significant effect of the nationality of a company (Japan vs. USA) on manufacturing technology use.

Further, Schroder and Sohal (1999) found that Australian-owned companies rate the anticipated benefits of increased throughput, sales, and investment in manufacturing technology more highly than foreign-owned companies from South Korea, Taiwan, Japan, USA, and New Zealand operating in Australia.

Unions

It has been widely suggested that effective implementing of manufacturing technology depends on the human factor or employees and their flexibility (Goldhar and Lei, 1994; Upton, 1995; Lefebvre et al, 1996). This often means that labor unions have to set aside their traditional work rules and job control strategies to allow team work and consultation (Osterman, 1994). Successful adoption of manufacturing technology also requires worker to attain new levels of operational skills and a higher level of commitment to improve product quality (Osterman, 1994). This can often be achieved through agreement with the union and management as in the case of Harley-Davidson Motor Company.

Chen et al, (1996) note that a company equipped with all the computerized or automated manufacturing technologies may be surprised to find that ultimate success is largely determined by the human factor. They also give the example of a plant, operated with the help of 300 robots, which had higher productivity and poorer quality performance than a more labor-intensive plant with a labor union.

Other major issue related to the adoption and implementation of manufacturing technology is employee commitment and cooperation (Krajewski and Ritzman, 1993; Chen and Gupta, 1993). Tchijov (1989) reports that plants with labor union membership exhibit the resistance to the adoption of manufacturing technologies. On the contrary, Dimnik and Richardson (1989) found that there was no relationship between union membership and adoption of manufacturing technology in a sample of auto-parts manufacturers in Canada. Small and Yasin (2000) investigated human factors in the adoption and performance of manufacturing technology in unionized organizations. They found a union effect on the adoption of just-in-time production system only. For all other technologies investigated in their study, there was no significant union effect. Thus, given the above, there is no clear evidence of union effect on manufacturing technology use; it deserves more investigation.

Performance Measures

Performance measures are multidimensional. Several researchers have investigated the relationship between manufacturing technology implementation and performance (Paul and Suresh, 1991; Chen and Small, 1994; Small and Yasin, 1997; Small, 1999; Swamidass and Kotha, 1998). This study classifies the wide range of performance measures in the literature into three groups:

- strategic measures
- organizational measures and

* business and market performance measures.

Strategic measures Researchers suggest that the performance measures of manufacturing technology implementation should be strategically focused (Millen and Sohal, 1998; Sohal, 1999; Efstathiades et al, 2000; Sun, 2000). These measures include many dimensions including quality and flexibility.

Quality has surfaced in many performance measures. For example, Dimnik and Richardson (1989) note that the key performance measures in evaluating manufacturing technology in the automotive industry in Canada are cost, quality and flexibility. Other researchers recommend other two dimensions while investigating the auto industry; product quality, and service quality comprising both pre- and after-sale service (Curkovic et al, 2000). In the literature this study find that quality performance measure may incorporate percent defective, rejection rate, customer complaints, and product accuracy (Paul and Suresh, 1991; Laosirihongthong and Paul, 2000).

Flexibility is an important component of performance especially in the automotive industry (Zairi, 1992; Zammuto & O'Connor, 1992; Sohal, 1994; Boyer, 1996). Small and Chen (1997) define flexibility as the ability to respond quickly to changing customer needs. They also classify manufacturing flexibility into two dimensions, "time-based flexibility" which focuses on the speed of response to customer needs, and "range-based flexibility" which is concerned with the ability to meet varying customization and volume requirements in a cost-effective manner. In addition, time-based performance of automotive suppliers is critical, and manufacturing lead-time is especially critical in this industry (Jayaram et al, 1999).

Organizational Performance

The specific measures of organizational performance include the degree to which manufacturing technology have improved work standard, skills of employees, image of the company, and coordination and communication within the company (Millen and Sohal, 1998; Sun, 2000; Efstahiades ct al, 2000). Organizational measures are related to workflow, work standardization, communication, and management control (Dean et al, 2000).

Business and Market Performance

A third set of measures is reported by Small and Yasin (1997), who suggest that business and market performance measures could be tied to revenue from manufacturing operation, return on investment, overhead cost, time-to-market for a new product, and market share of existing/new products. Some of these measures are financial performance measures. Swamidass and Kotha (1998) investigated the relationship between manufacturing technology use and financial performance. They found that the relationship is not significant, and conclude that perhaps strategic rather than financial benefits might have been the primary reason for investing in manufacturing technology. Therefore, this study did not use financial performance measure.

In summary, performance measures used in manufacturing and operations management researches while investigating manufacturing technology use are varied. However, there is a common understanding that there are three important but broad dimensions of performance measures -- quality, flexibility, and organizational measures. This study uses these three dimensions for performance measurement reflecting the successful for manufacturing technology implementation.

Guiding Research Question

The discussion of key variables and their relationships above provide the basis for the guiding research question of the study based on the three technology types and three performance dimensions discussed above: Whether High-volume automation technologies, data-interchange technologies, and low-volume automation technologies, either individually or collectively affect one or more of the performance measures, which are quality performance, flexibility and organizational performance.

RESEARCH METHODOLOGY AND DATA ANALYSIS

Sample and Data Collection

This study selected only companies who are listed with Thailand Industrial Standard Institute and Thai Automotive Institute. The companies surveyed in this study all produce products classified in the automobile and parts/components industry sector. Questionnaire used in this study consists of three parts: the degree of manufacturing technology use, perceived manufacturing technology benefits/performances, and organizational characteristics. It includes fifteen manufacturing technology (Boyer et al., 1997; Burgess and Gules, 1998; Efstahiades et al., 2000; Boyer and Pagell, 2000; Efstathiades et al., 2002), thirteen perceived performance measures (Small and Yasin, 1997; Park, 2000), and four organizational characteristics including size of the company (measured by a number of employees), type of ownership, and existence of labor union.

Table 1: Characteristics of respondents 1 Size classification according to Ministry of Industry, Thailand

Characteristics	Description	%
Respondents	MD/VP/P	10.20
	Factory/Production Mgr.	37.80
	General Manager	14.50
	Engineering Mgr.	22.70
	QA/QC Mgr.	18.80
Company size (number of employees)[1]	Small to medium <= 200	58.40
	Large > 200	41.60
Ownership	Thai-owned	30.40
	Foreign-owned	14.30
	Joint-venture	55.30
Labor union	Labor union present	30.45
	No labor union	69.55
Main product classifications	Body parts	21.42
	Chassis parts	25.58
	Suspensions parts	12.25
	Electrical parts	8.20
	Accessories	11.45
	Trim parts	21.10
Existing quality management system	ISO/QS9000 certified	94.58
	None	5.42

Totals of 480 questionnaires this study distributed to factory, general, engineering, and quality assurance managers who have a responsibility for manufacturing technology implementation in their own companies. Questionnaires were sent to the respondents by given directly (for return by mail) at the suppliers' monthly meeting of one Japanese assembler and one American assembler. One respondent per company was asked to indicate the degree of implementation for fifteen manufacturing technology and perceived performance after the implementation. The usage attributed to these technologies and performances was measured using Likert's five-point scale where 1 = not used ornot satisfied and 5= extensively used or very satisfied. A total of 124 questionnaires were returned giving a response rate of 25.83 percent, comparable to the rates in previous such research (Sohal, 1996; Small and Chen, 1997). Table I exhibits the characteristics of respondents.

Non-Respondent Bias

A random sample of 30 companies from the 356 non-respondents was selected to compare the respondents with non-respondents. The following classificatory data this study are collected from the 30 non-respondents through the phone: (1) size (employment), (2) ownership, (3) ISO 9000 certification, and (4) unionization. All 30 non-respondents contacted by phone provided classificatory information requested by phone. In Table II, this study indicates the result of the comparison between responding and non-responding sample. The Chi-square values indicate that the two samples are statistically different. Major differences between respondents and non-respondents being that the sample of respondents have larger firms, foreign-owned firms, more ISO-certified firms, and more unionized forms. If this study assume that the sample of 30 non-respondents is representative of all non-respondents, the findings of this study are pertinent to the 124 manufacturers who participated in this study.

Table 2: Comparison of Respondents with a Random Sample of Non-Respondents. * The Chisquared values for size, ownership, ISO certification and union are all larger than the Chisquare table values for .05 significance (2-tail). Thus, the respondents are not similar to the random sample of non-respondents

Organizational characteristics	Respondents	Non-respondents	Chi-sq.	Chi-Sq. table (.05 significance, 2-tail)*
Size =< 200 employees	72(58%)	12(40%)		4.89
Size > 200 employees	52(42%)	18(60%)	17.1	
Thai owned	37(30%)	6(20%)		7.57
Foreign owned	17(13.7%)	15(50%)		
Joint venture	70(56.3%)	9(30%)	52.8	
ISO/QS9000 certified	117(94.4%)	19(63.4%)		4.97
None	7(5.6%)	11(36.6%)	8.2	
Labor union present	37(30%)	10(33.4%)		5.14
No labor union	87(70%)	20(66.6%)	10.4	

Generalization of the findings to non-respondents must be done with care. Given that the sample of 124 firms participating in this study is substantial, the findings are valuable even if they are not representative of the entire Thai auto industry.

Data Analysis

The Reliability and Validity of Empirical Measures

The internal consistency of our measures was verified using Cronbach's alpha (Cronbach, 1951); a value greater than 0.6 was treated acceptable (Chen and Small, 1994). Content validity was established from literature review, expert and practitioner opinions, and pre-testing with a small number of managers. Construct validity was ensured by factor identification through principal component factor analysis (Nunnally, 1967). Factors are selected using these three rules: (a) minimum Eigenvalue of 1, or cumulative factor variance explained in excess of 70 percent; (b) minimum factor loading of 0.4 for each item; and (c) the simplicity of factor structure. Factor analysis was used to find factors to explain dependent variables (performance measures) and independent variables (technology use). SPSS software was used to perform principal component analysis including an orthogonal transformation with Varimax rotation. The results are shown in Tables III (for technology factors) and VII (for performance factors). In order to test the validity of perceptual performance measures, this study conducted a correlation analysis between selected objective external measures with self-reported perceptual data on performance for 20 per cent of the companies randomly selected (n = 30) from our sample of 124 respondents. Selected objective external measures were obtained from the Monthly Suppliers Evaluation Reports--MSER (Sriwatana, 2000; Vibulsilapa, 2000) concerning delivery, quality, cost, and organizational reputation. Correlation analysis between MSER data and survey data was conducted, specifically, the correlation analysis between MSER data and survey-based composite values of flexibility, quality performance, and organizational performance for a random sample of 30 companies. The resulting correlation coefficients are 0.77, 0.81, and 0.73 respectively. Therefore, this study considers the perceptual performance measures acceptable (Swamidass and Kotha, 1998; Lewis and Boyer, 2002).

RESEARCH FINDINGS

Technology use (factors) Confirm Prior Studies

Multi-item scales are developed for each construct (technology and performance) in this study. Before creating the final scales, the data are checked for normality and outliners. As shown in Table III and VII, the Kaiser-MeyerOlkin (KMO) measure of sampling adequacy is 0.887 (for technology factors) and 0.894 (for performance). A minimum Kaiser-Meyer-Olkin score of 0.5 is considered necessary to reliably use factor analysis for data analysis (Small, 1999). Score over 0.80 are considered very strong. Similarly, the Bartlett test of sphericity (the higher, the better) was 987.32 (technology factor) and 1322.t (performance) with significance value (Small, 1999).

The results of rotated principal component factor analysis show that three factors explain 63.25 per cent of the total variance (Table III). These technology factors are used in subsequent analysis to examine the relationships between technology use and organizational characteristics, as well as technology use and performance. In Table III, the result indicates that seven technologies load on the first factor. This factor consists of technologies that can be used to reduce direct labor costs in repetitive operations and high-volume production with low variety of products. Therefore, the study names this factor as "High-volume automation technologies." The second factor consists of five technologies that relate to planning and data interchange. Therefore, the study names this factor as "Data-interchange technologies," which parallels the "information exchange and planning technologies" reported by Swamidass and Kotha (1998) using US data. The third factor includes technologies that provide low-volume manufacturing flexibility that permits low-volume high variety production. This study, therefore, calls this factor, "Low-volume flexible automation technologies." The three factors that emerged from data of the Thai automotive industry are similar to technology factors that determined from factor analysis of some previous studies. Thus, it is important to note that manufacturing technology factors that were identified in this study are robust and are stable across time and national boundaries.

Table 3: Technology Facts (Rotated Comonent Matrix). Note: 1 – Lowest, 5 - Highest

Technology Factors	Mean	S.D.	orsExtracted fact		
			1	2	3
High-volume automation technologies					
Automated material handling	2.15	1.21	0.774		
Automated assembly system	2.38	1.35	0.732		
Automated storage/retrieval system	1.74	1.02	0.715		
Automated inspection system	2.42	1.01	0.701		
Computer-aided manufacturing	3.15	1.04	0.554		
Barcode system	2.88	1.33	0.568		
Pick and place robots	2.04	1.41	0.520		
Average mean score	2.39				
Data interchange technologies					
Material resources planning	2.54	1.11		0.726	
cturing centerFlexible manufa	1.98	1.06		0.711	
Computer-aided process planning	2.22	1.21		0.702	
Computerized statistical process control	2.16	1.05		0.566	
Electronic data interchange	2.53	1.02		0.511	
Average mean score	2.28				
Low-volume flexible automation technologies					
Computer numerical control	3.88	1.44			0.818
Pneumatic and hydraulic equipment	3.72	1.32			0.735
Computer-aided design	3.25	1.51			0.598
Average mean score	3.62				
Kaiser-Meyer Olkin adequacy(KMO)			0.887		
ericityBartlett's test of sph			987.32		
Significance			0.00000		
Cronbach's Alpha			0.875	0.902	0.821
Eigenvalues			3.488	2.876	2.034
Varience explained			24.45	22.62	16.18
Total variance explained			24.45	47.04	63.25

Technology Factors and Organizational Characteristics

Size

In Table IV, this study compares the use of three different technology dimensions (factors) in large versus small/medium firms. The table shows that there is a significant difference between large and small-to-medium companies in the use of High-volume automation technologies (p=.025) and Low-volume flexible automation technologies (p=.002). There is no significant difference in the use of Data-interchange technologies (p=.103). Data-interchange technologies form the backbone of manufacturing systems now and these technologies have been around longer the other technologies. The implication is that all manufacturers, regardless of size, equally depend on Data-interchange technologies. One reason being, these technologies are easily implementable on PCs, which are affordable by even small manufacturers. For example, MRP and Electronic Data Interchange (EDI) (see Table III) that are included in this dimension could be implemented using ordinary PCs. The findings reveal that plant size has differential effect on the various technology factors.

Table 4: Technology Factors and Size of Company. * (Employees <= 200 = small-tomedium; employees > 200 = large.)

Technology Factors	Sig.	Small-to-medium Composite mean	Large Composite mean
High-volume automation technologies	0.025*	2.87	2.74
Data-interchange technologies	0.103	2.01	2.23
Low-volume flexible automation technologies	0.002**	2.66	3.37

* Significant at 0.10 level. ** Significant at 0.05 level. *** Significant at 0.01 level.

Ownership

Table V reports the use of the three different dimensions of manufacturing technologies in Thai-owned, foreign-owned and jointly-owned firms. According to the table, the following is revealed:

- In foreign-owned plants, High-volume automation technology use is significantly higher than its use in either Thai-owned (p=.001) or joint-venture plants (p=.001).

- In Thai-owned plants, Low-volume flexible automation technology use is higher than the use of this technology in either joint ventures (p=.001) or foreignowned (p=.001) plants. Apparently, Thai plants produce more low volume components.

• Plant ownership has no effect on Data-interchange technologies. In an earlier section, this study reported that plant size has no effect on Data-interchange technology use. Taken together with this finding, it is important to note that Data-interchange technologies are relatively more mature technologies, easily implementable without much capital or resources, and is immune to size and ownership.

Table 5: Technology Factors and Ownership. * Significant at 0.10 level. ** Significant at 0.05 level. *** Significant at 0.01 level. ns = not significant

Technology Factors		Thai-owned	Joint-venture	Foreign-owned
High-volume automation technologies	Mean score →	2.35	2.22	2.61
Significance of Joint venture and column	Joint venture	p=0.182 (ns)		
Significance of Foreign-owned and column	Foreign-owned	p=0.001***	p=0.001***	
Data-interchange technologies	Mean score →	2.53	2.72	2.45
Significance of Joint venture and column	Joint venture	p=0.225 (ns)		
Significance of Foreign-owned and column	Foreign-owned	p=0.743 (ns)	p=0.351 (ns)	
Low-volume flexible automation technologies	Mean score →	3.47	3.18	3.01
Significance of Joint venture and column	Joint venture	p=0.001***		
Significance of Foreign-owned and column	Foreign-owned	p=0.001***	p=0.423 (ns)	

Unionization

Very few studies have investigated the effect of unionization on manufacturing technology use. Tchijov (1989)'s found that plants with labor union membership exhibit the resistance to adoption of new technologies. This study does not measure union membership of employees, if measures if the plant is unionized or not. As shown in Table VI, the use of Data interchange technologies is significantly higher (p=.013) in plants with labor unions, and the use of Highvolume automation technologies is higher in non-union plants (p=.011). It is a notable finding that unionization does have an effect in the use of at least a certain technology.

Table 6: Technology Factors and Labor Unionization. * Significant at 0.10 level. ** Significant at 0.05 level. *** Significant at 0.01 level.

Technology Factors	Sig.	Labor union Composite mean	Non-union Composite mean
High-volume automation technologies	p=0.011*	2.53	2.62
Data-interchange technologies	p=.013**	2.77	2.32
Low-volume flexible automation technologies	p=0.644	3.32	3.15

Performance Measures

A principal component factor analysis is used to reduce and group the thirteen individual performance items in the survey into three performance factors, "Flexibility performance", "Quality performance", and "Organizational performance". The three performance factors together explain 71.55 percent of the total variance (Table VII).

Technology Factors and Performance

As a rule, this study finds that there is little association between technology use and performance factors (Table VIII), the one exception being High-volume automation technology, which is associated with Quality Performance (Pearson r =0.236; p = 0.000). Three multiple regression models to estimate performance using technology use dimensions are reported in Table IV. According to the table, only quality performance is explained by one of the technology dimensions (High-volume automation technologies). An inference from this study is that, for the auto industry, high-volume automation is an essential ingredient for quality. This inference may be limited to the auto industry because of the sample.

Table 7: Performance Factors (Rotated Component Matrix Note: 1 – Lowest, 5 – Highest

Performance measures	*Mean	S.D.	Extracted Factors		
			1	2	3
Flexibility performance					
Delivery lead time	3.87	0.84	0.720		
Responsiveness to customer needs	3.65	0.78	0.815		
Production change overtime	3.42	0.92	0.736		
Set-up time	3.33	0.76	0.884		
Average mean score	3.57				
Quality performance					
Defective ratio along the process	3.66	0.88		0.833	
Rejection ratio within the process	3.47	0.91		0.784	
Customer complain	4.22	1.02		0.746	
Frequency of inspection	3.85	0.77		0.626	
Accuracy of product	4.01	0.98		0.689	
Average mean score	3.84				
Organizational performance					
Upgrading human skills	3.72	0.74			0.843
Company's image	3.88	0.83			0.744
Work standardization	4.21	0,98			0.832
Reducing bargaining of skilled labor	3.18	0.86			0.675
Average mean score	3.75				
Kaiser-Meyer Olkin adequacy(KMO)			0.894		
Bartlett's test of sphericity			1322.7		
Significance			0.00000		
Cronbach's Alpha			0.922	0.916	0.842
Eigenvalues			2.133	3.411	2.756
Varience explained			24.22	28.72	18.61
Total varience explained			24.22	52.94	71.55

Table 8: Correlation Analysis between Technology Factors and Performance Factors. * Significant at 0.10 level. ** Significant at 0.05 level. *** Significant at 0.01 level

Technology Factors	Flexibility	Quality	Organizational
High-volume automation technologies	0.005 p = 0.843	0.236 p = 0.000***	0.054 p = 0.331
Data-interchange technologies	0.054 p = 0.466	0.082 p = 0.342	0.037 p = 0.693
High-flexible automation technologies	0.993 p = 0.215	0.051 p = 0.442	0.027 p = 0.578

Table 9: Technology Factors and Size of Company* * Employees <− 200 − small-tomedium; employees > 200 = large. * Significant at 0.10 level. ** Significant at 0.05 level. *** Significant at 0.01 level

Technology Factors	Sig.	Small-to-medium Composite mean	Large Composite mean
High-volume automation technologies	0.025*	2.87	2.74
Data-interchange technologies	0.103	2.01	2.23
Low-volume flexible automation technologies	0.002*	2.66	3.37

CONCLUSIONS AND FUTURE STUDIES

The most notable theme here is that findings from this study confirm several findings reported in the literature based on data from other nations. First, the study concurs with previous studies that show the size of companies influences the use of manufacturing technology. The reasoning is now this study known; large companies can afford the higher cost of adopting these technologies. Also, managerial resources necessary in planning and implementing such technologies are available in larger companies (Ariss et al, 2000).

Second, this study found that technology use is a function of the nationality of the plant ownership. For example, finding indicates that High-volume automation technologies such as automated material handling, automated assemblysystem and robots are more likely to be adopted in foreign-owned companies than in Thai-owned and joint-venture companies. Foreign-owned companies perhaps tend to adopt more technologies because of their superior financial, technical and managerial resources, technological capabilities, and abilities to transfer those technologies. Further, foreign-owned plants may replicate the use of technology in plants back home, which is invariably a more developed nation compared to Thailand. The findings concerning the effect of the nationality of ownership on technology use concurs with studies on technology implementation in Australia (Sohal et al, 1991), in the UK (Sohal, 1994), and the USA (Kotha and Swamidass, 1998).

Third, The multidimensional view of technology reported by Swamidass and Kotha (1998) using a US sample holds up this study in the sample of firms from Thai auto industry; further, the two samples are several years apart.

Some Directions for Future Studies

The Need for More Investigations of the Unionization-Technology Link

A notable finding of this study is that the use of Data interchange technologies, at least, is significantly higher in plants with labor unions. Could it be that these technologies reduce the influence or soften the effect of unionization? Do they reduce the need for employees in functions affected by unionization? Is it possible that unions do not resist the adoption of Data-interchange technologies? The search for answering to these questions is a worthy line of investigation for the future.

A Proposed Concept of Manufacturing Technology Use

This study, confirms the emerging multi-dimensional view of technology use with collected data in Thailand with a specific industry. Further, the multiple technology factors that this study found in Thailand are similar to those found in the USA. This is a testimony to the robustness of the technology factors, which transcend national borders. Additionally, in an earlier study by Swamidass and Kotha (1998), which reported the multiple dimensions of technology, the data came from a survey nearly 10 years earlier than the Thai survey reported here. Therefore, it appears that the technology dimensions/factors are stable across time.

In addition, this study confirms findings concerning the effect of plant size, and the nationality of ownership. Taken together, empirical research to this point encourages the following Theory of manufacturing technology use for testing and retesting in the future for its confirmation and establishment: " In the complex manufacturing environment made of people, technology and procedures, manufacturing technology is not homogenous but has consistently distinct dimensions. These technology dimensions are robust and exist across national boundaries and time. However, technology use is a function of plant size, and the nationality of plant ownership".

Limitations

While this study is based on responses from nearly 150 firms, our nonresponse bias test shows that the responding firms are larger, more foreignowned, more ISO-certified, and more unionized, compared to nonrespondents. In the future, a more representative sample may be investigated. Boyer et al (1997) found that companies benefit from manufacturing technology investments when there is adequate and matching investments in the infrastructure. This study did not investigate this aspect of technology use in more details. Therefore, this study would encourage studies that test the above concept in order to expand it to cover the role of infrastructure investments.

ACKNOWLEDGEMENTS

The author would like to thank Paul M. Swamidass, Professor and Director of Thomas Walter Center of Technology, Auburn University, Alabama, for his valuable suggestions on the first revision of this manuscript.

REFERENCES

1. Adler, P.S. (1988). "Managing flexible automation". California Management Review. 20 (1), 35-56.

2. Ariss, S.S., Raghunathan T.T. and Kunnathar A. (2000). "Factors affecting the adoption of advanced manufacturing technology in small firms". S.A.M. Advanced Management Journal, Spring.

3. Attasathavorn, J. (2001). "Reports of Thai Automotive Industry". For Quality Magazine, March-April, 36-49. (in Thai).

4. Bank of Thailand. (2000). The Economics Report during January – March 2001, 65- 80. (in Thai).

5. Boer, H., Hill, M. and Krabbendam, K. (1990). "FMS implementation management: promise and performance". International Journal of Operations and Production Management, 10 (1), 5-20.

6. Board of Investment (BOI). (1995). Report of the Investment of Automotive Industry in Thailand. Board of Investment, Bangkok. (in Thai).

7. Boyer, K. (1996). "An assessment of managerial commitment to lean production". International Journal of Operations and Production Management, 16 (9), 48-59.

8. Boyer, K, Leong, G.K., Ward, P.T., and Krajewski, L.J. (1997). "Unlocking the potential of advanced manufacturing technologies". Journal of Operations Management, 15, 331-347.

9. Chen, I.J., Gupta A., and Chung, C.H. (1996). "Employee commitment to the implementation of flexible manufacturing systems". International Journal of Operations and Production Management, 16 (7), 4-13.

10. Chen, I.J. and Gupta, A. (1993). .Understanding the human aspect of flexible manufacturing system through management development.. International Journal of Management Development, 10 (1), 32-43.

11. Chen, I.J. and Small M.H. (1994). "Implementing advanced manufacturing technology: An Integrated Planning Model". OMEGA, 22 (1), 91-103.

12. Cronbach, L.J. (1951). Coefficient Alpha and the Internal Structure of Tests: Psychometrika, 16, 297-334.

13. Curkovic, S., Vickery, S.K., and Droge, C. (2000). "An empirical of the competitive dimensions of quality performance in the Automotive supply industry". International Journal of Operations and Production Management, 20 (3), 386- 403.

14. Dean, A.S., Mcdermott, C., Stock, G.N. (2000). "Advanced manufacturing technology: Does more radicalness mean more perceived benefits?". The Journal of High Technology Management Research, 11(1), 19-33.

15. Dean, J.W. Jr. and Snell, S.A. (1991). .Integrated manufacturing and job design.. Academy of Management Journal, 34 (4), 776-804.

16. Dean, J.W. Jr., Yoon, S.J., and Susman, G.I. (1992). ."Advance manufacturing technology and organizational structure: Empowerment or subordination?. Organizational Science, 3 (2), 203-229.

17. Dimnik, T. and Richardson, R. (1989). "Flexible automation in the auto parts industry". Business Quarterly, 54 (4), 46-53.

18. Efstathiades, A., Tassou A.S., Oxinos G., Antoniou A. (2000). "Advanced manufacturing technology transfer and implementation in developing countries: The case of the Cypriot manufacturing industry". Technovation, (2), 93-102.

19. Ettlie, J.E. (1984). Implementation strategy for discrete parts manufacturing innovation. In: Warner, M. (Ed.), In Microelectronics, Bookfield, VT.

20. Federation of Thai Industries (FTI). (2000). Reports of The Thai Automotive Industry. Working group of automotive industry, Bangkok. (in Thai).

21. Japan International Corporation Agency (JICA). (1995). The Study on Industrial Sector Development Supporting Industries in the Kingdom of Thailand. Tokyo: International Corporation. (in Thai).

22. Jayaram, J., Vickery, S.K. and Droge, C. (1999). "An empirical study of time-based competition in the North American automobile supplier industry". International Journal of Operations and Production Management, 19 (10), 1010- 1033.

23. Kotha, S. and Swamidass, P.M. (1998). "Advanced manufacturing technology use: exploring the effect of the nationality variable". International Journal of Production Research, 36 (11), 3135-3146.

24. Krajewski, L.J., and Ritzman, L.P. (1993). Operation Management: Strategy and analysis. 3rd Edition: Addison, Reading, MA.

25. Laosirihongthong, T. and Paul, H. (2000). "Implementation of New Manufacturing Technology and Quality Management System in Thai Automotive Industry". Proceedings of IEEE International Conference in Management and Innovation of Technology, November 12-15, 2000, Singapore.

26. Lefley, F. and Sarkis, J. (1997). "Short-termism and the appraisal of AMT capital projects in the USA and UK". International Journal of Production Research, 35 (2), 341-369.

27. Mansfield, E. (1993). "The diffusion of flexible manufacturing system in Japan, Europe and the United States". Management Science, 39 (2), 149-159.

28. Meredith, J. R. (1987). "Implementing new manufacturing technologies: managerial lessons over the FMS life cycle". Interface, November-December, 51-62.

29. Millen, R. and Sohal A.S. (1998). "Planning processes for advanced manufacturing technology by large American manufacturers". Technovation, 18 (12), 741-50.

30. Nunnally, J.C. (1967). Psychometric Theory, McGraw Hill: New York, NY Park, Y.T. (2000). "National systems of Advanced Manufacturing Technology (AMT): Hierarchical classification scheme and policy formulation process". Technovation, (20), 151-159.

31. Paul, H. and Suresh B. (1991). "Manufacturing strategy through planning and control techniques of advanced manufacturing technology". International Journal of Technology Management, 6 (3-4), 233-242.

32. Paul, H. and Laosirihongthong T. (1999). ISO9000 Implementation in Thailand: Experience form Thai Autoparts Industry. Proceedings of the 14th International Conference in CAD/CAM, Robotics and Factory in the Future, Narosa Publishing House, 527-532.

33. Peter, B., Lee, G., and Sohal, A.S. (1999). Lessons for implementing "AMT: Some case experience with CNC in Australia, Britain and Canada". International Journal of Operation and Production Management, 19 (5/6), 515-526.

34. Rosenthal, S.R. (1984). "Progress toward the 'factory of the future". Journal of Operation Management, 4 (3), 405-415.

35. Schroder, R. and Sohal A.S. (1999). "Organizational characteristics associated with AMT adoption: Towards a contingency framework".

International Journal of Operations and Production Management, 19 (12), 1270-1291.

36. Small, M.H. (1999). "Assessing manufacturing performance: an advanced manufacturing technology portfolio perspective". Industrial Management & Data Systems, 99(6), 266-277.

37. Small, M.H. and Chen I.J. (1997). "Organizational development and time-based flexibility: An empirical analysis of AMT adoption". International Journal of Production Research, 35 (11), 3005-3021.

38. Small, M. H. and Yasin M.M. (1997). "Developing a framework for the effective planning and implementation of advanced manufacturing technology". International Journal of Operations and Production Management, 17 (5), 468-489.

39. Sohal, A.S. (1994). "Investing in advanced manufacturing technology: Comparing Australia and the United Kingdom". Benchmarking for Quality Management & Technology, 1 (1), 24-41.

40. Sohal, A.S. (1999). "Introducing New Technology into a Small Business: A Case Study of Australia Manufacturers". Technovation, 19 (3), 187-193.

41. Sohal, A.S., Samson, D. and Weill, P. (1991).,"Manufacturing and technology strategy: a survey of planning for MANUFACTURING TECHNOLOGY". Computer Integrated Manufacturing System, 4, 71-79.

42. Sriwatana, T. (2000). Summary of suppliers performance evaluation. The Monthly Suppliers Evaluation Reports, 1998-2000. Toyota Motor (Thailand) Company Limited. 45-70.

43. Sun, H. (2000). "Current and future patterns of using advanced manufacturing technologies". Technovation, (20), 631-641.

44. Swamidass, P.M. (2000). Encyclopedia of Production and Manufacturing Management, Kluthis studyr Academic Publishers. 400-405.

45. Swamidass, P.M. and Kotha S. (1998). "Explaining manufacturing technology use, firm size and performance using a multidimensional view of technology". Journal of Operation Management, 17, 23-37.

46. Tchijov, I. (1989). "CIM Introduction: Some Socioeconomic Aspects". Technological Forecasting and Social Change, Vol. 35 (2-3), 261-275.

47. Thai Automotive Institute (TAI). (2000). Thailand Automotive Industry Directory 2000. Bangkok. (in Thai).

48. Thailand Development Research Institute (TDRI). (1999). The development of Thailand's Technological Capability in Industry,

Bangkok. (in Thai).

49. Vibulsilapa, S. (2000). Suppliers evaluation results. Quality Assurance Supplier Evaluation Reports, 1998-2000. Isuzu Motor (Thailand) Company Limited, 81-124.

50. Warner, T. (1987). "Information technology as a competitive burden". Sloan Management Review, Fall, 55-61.

51. Zairi, M. (1992). "Measuring success in AMT implementation using customersupplier interaction criteria". International Journal of Operations and Production Management, 12 (10), 34-55.

52. Zammuto, R.F. and O'Connor, E.J. (1992)."Gaining advanced manufacturing technology benefits: The roles of organization design and culture". Academy of Management Review, 17, 701-728.

Chapter 12

SELECTION OF ADDITIVE MANUFACTURING TECHNOLOGIES USING DECISION METHODS

Anderson Vicente Borille and Jefferson de Oliveira Gomes

Technological Institute of Aeronautics - ITA Brazil

INTRODUCTION

The use of Rapid Prototyping technologies is becoming increasingly popular due to the reduction of machinery prices. Consequently, more and more industries now have the opportunity to apply such processes to improve their product development cycles. The term Rapid Prototyping was commercially introduced to highlight the first application, the quickly production of prototypes into the product development process. Improvements were done in the quality of the equipments and the variety of materials. Furthermore, new processes were introduced into the market, which enlarged the application's range of Rapid Prototyping technologies. As a consequence, new terms were also used to describe the final application of such technologies as Rapid Manufacturing (RM); Rapid Tooling (RT), which indicates the use of such technologies to produce moulds and tooling, etc.

However, as important as to identify the technical limits of the each technology, it is needed to balance the characteristics of each process in order to decide which one fulfills the product requirements the best way. And this should be done systematically using a decision method. The decision method, in turn, should be able to evaluate the relative weights of product requirements related to the process capabilities. It is not just a matter of manufacturing process substitution. It is possible – and desirable in case of RM – to modify designing and product development processes too. This chapter is divided into two sections. The first part considers prototyping applications, where the requirements of the part to be produced are not too severe. In this case, available process capabilities should be used to satisfy costumer's needs, usually at the lowest manufacturing cost and delivery time possible. The second section is intended to those who are concerned in Rapid Manufacturing Applications. Rapid Manufacturing means that the parts will be produced as end product, thus, the product requirements are more rigorous then prototyping applications.

PART I: RAPID PROTOTYPING APPLICATIONS

This chapter aims to present different decision making approaches to choose an adequate RP process. Here, four decision approaches were applied to compare six processes regarding six criteria, using the input data from previous works. As result, three decision methods were compared, additionally to the references. Two different scenarios were constructed, where different important attributes were considered, simulating two differentprototype applications. It was demonstrated that not all methods result to the same RP ranking, however most of them provide the same first option for a given scenario. The characteristics of the methods could be related to their influence on the evaluation, which serve as guidelines for the decision makers in order to reflect their exact opinion or requirements. Although the fundamentals of the decision methods are presented here, one should be careful while comparing the RP process, because their attributes may vary enormously depending on the parameter process to build a part. Despite all the considerations and precautions to be observed, the selection of the RP process can be done in a simple way, dispensing complex calculations.

Example of Application

The decision process requires the evaluation of alternative characteristics (attributes) regarding the desired requirements (criteria) to reach an objective. Byun and Lee (2005), based on questionnaires answered by users, concluded that the following six attributes are the most important regarding the use of RP processes: accuracy (A), surface roughness (R), tensile strength (E), elongation (S), cost of the part (C) and build time (B). Further, they gathered these attributes from six different RP processes, and proposed a method to evaluate these attributes simulating two different scenarios: Scenario 1) where the cost of the part (C) and build time (B) were considered most important factors, followed by S and E, and A and R, and Scenario 2) where accuracy (A) and surface roughness (S) where considered most important followed by S and E, and C and B. Later, Padmanabhan (2007) used the same RP processes attributes to evaluate similar conditions, but using Graph Theory & Matrix Approach (GT&MA) instead of Topsis. The attributes of the Alternatives presented in Table 1 were used by both previous works.

Table 1: Alternatives attributes table (Byun and Lee, 2005; Rao and Padmanabhan, 2007)

Process	A	R	S	E	C	B
Process1	120	6,5	65	5	Very high	Medium
Process2	150	12,5	40	8,5	Very high	Medium
Process3	125	21	30	10	High	Very high
Process4	185	20	25	10	Slightly high	Slightly low
Process5	95	3,5	30	6	Very high	Slightly low
Process6	600	15,5	5	1	Very very low	Very low

Based on the information from the processes and from the requirements, a decision maker should be able to evaluate the alternatives and propose a recommendation. The issues to manage consist that most product requirements are contradictory. For example, in the Table 1 the process which has the lowest cost produces the weakest part. The decision maker should be able to answer – in a systematically form – how much more important is the cost in relation to tensile strength? Such questions are well complicated to be translated into numbers directly, but using established procedures the answer can be very consistent.

Decision Making processes are usually elaborated to be useful to a large range of applications, consequently, they have to be lapidated to be applied to each specific use. An important point of this work, is that for each decision approach, some kind of consideration had to be done in order to represent an approximated scenario to different decisionmethods. They were most related to the conversion of scales and weighting procedures. Even with these considerations, most decision methods provided the same process as the first option. Thus, the decision maker may feel free to use the most familiar way, just considering some rough characteristics.

Analytic Hierarchy Process (AHP)

The Analytic Hierarchy Process (AHP) is a multi-criteria decision-making approach and was introduced by Thomas L. Saaty (Saaty, 1977; Saaty 1990). The AHP has attracted the interest of many researchers mainly due to the mathematical properties of the method and the fact that the required input data is rather easy to obtain (Triantaphyllou, 1995, Guglielmetti et. al. 2003).

a) Method

The method is based on a pairwise comparison of alternatives and criteria of a hierarchical structure (Fig. 1). In order to evaluate the approach, a comparison matrix for the criteria must be described, as the Fig. 2.

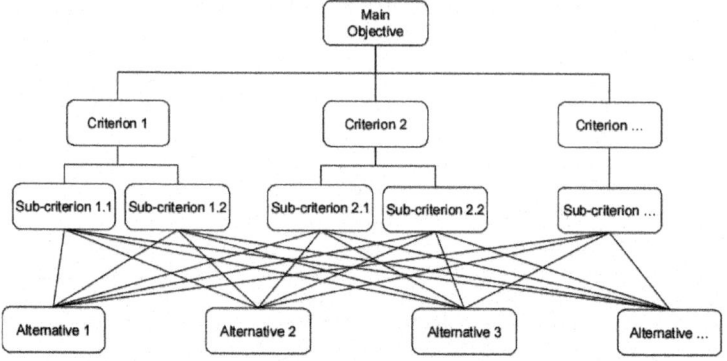

Figure 1: The hierarchical structure of AHP approach (Saaty, 1977).

$$\begin{array}{cccc} & C_1 & C_2 & \cdots & C_n \end{array}$$

$$\begin{array}{c} C_1 \\ C_2 \\ \cdot\cdot \\ C_n \end{array} \begin{pmatrix} w_1/w_1 & w_1/w_2 & \cdots & w_1/w_n \\ w_2/w_1 & w_2/w_2 & \cdots & w_2/w_n \\ \cdots & \cdots & \cdots & \cdots \\ w_n/w_1 & w_n/w_2 & \cdots & w_n/w_n \end{pmatrix}$$

Figure 2: Comparison matrix fort the criteria (Saaty, 1977).

Each element w_i/w_j have to represent how much the i criteria is more important than the j, following the fundamental scale from Saaty (Table 2).

Intensity of importance	Definition	Explanation
1	Equal importance	Two activities contribute equally to the objective
3	Moderate importance of one over another	Experience and judgment slightly favour one activity over another
5	Essential or strong importance	Experience and judgment strongly favour one activity over another
7	Very strong importance	An activity is strongly favoured and its dominance demonstrated in practice
9	Extreme importance	The evidence favouring one activity over another is of the highest possible order of affirmation
2,4,6,8	Intermediate values between the two adjacent judgments	When compromise is needed
Reciprocals	If activity *i* has one of the above numbers assigned to it when compared to *j*, then *j* has the reciprocal value when compared with *i*.	
Rationals	Ratios arising from the scale	If consistency were to be forced by obtaining n numerical values to span the matrix

Table 2: The fundamental scale (Saaty, 1977)

In order to evaluate the criteria matrix using the AHP method, the principal eigenvector must be calculated. Saaty (2003) justified that the eigenvector has two meanings: first, is a numerical ranking of the alternatives, and second, the ordering should also reflect intensity as indicated by the ratios of the numerical values. The explanation of why the eigenvector should be used (Saaty, 2003; Saaty, 1977) as well how to calculate it (Saaty 2000) can be found in the respective literature.

The criteria matrix should be then evaluated related to consistency, because, despite their best efforts, people's feelings and preferences remain inconsistent and intransitive (Saaty, 1977). Although the AHP approach permits some inconsistency, Saaty accept the judgments w if the consistency ratio (CR) is less than 10%, where:

$$CR = \frac{CI}{RI} \tag{1}$$

$$CI = \frac{\lambda_{max} - n}{n - 1} \tag{2}$$

Where n is the order of the considered matrix, and RI (random index) given by Saaty (2000) (Table 3).

Table 3: Random index (Saaty, 2000)

n	1	2	3	4	5	6	7	8	9	10	11	12	13	14	15
RI	0	0	0,52	0,89	1,11	1,25	1,35	1,40	1,45	1,49	1,51	1,54	1,56	1,57	1,58

After evaluating the criteria matrix, the alternatives must be analysed, through the use of matrixes and calculations of the principal eigenvector, which, in turn, is a Column Matrix. For each criterion, a matrix similar to Fig. 1 must be built, but comparing all the alternatives, following the same weight considerations presented in Table 2. Consequently, n+1 matrices should be created, where n is the number of criteria – one criteria matrix and one matrix of the alternatives for each criterion. So, n eigenvectors are obtained from n alternative matrices (Column Matrix), which are combined into a new nxn matrix. This last matrix is then multiplied by the eigenvector of the criteria matrix. The final ranking of the alternatives results from this multiplication.

b) Application

Using the initial data of the attributes of RP processes presented in Table 1 and the relative importance of criteria in Scenario 1 and Scenario 2 described above, a decision maker is able to execute a process selection using AHP. The first step is to convert the qualitative and quantitative inputs from Table 1 into the fundamental scale of Saaty. Second, the criteria data (weights) must be also converted in the AHP matrix-format to calculate the local eigenvector.

In order to convert the qualitative analysis of cost (C) and build time (B) into numerical values, the results of the machines were compared pairwise to each other in a criteria matrix, and the eigenvector calculated to define local priorities. For this, initially, the 9 linguistic terms – very very low, very low, ..medium... very high, very very high – from Byun and Lee (2005) were converted into the numbers 1 through 9. So, a matrix of combinations could be built as the Table 4. Then, for each criteria (C and B), a matrix was built comparing the attributes of each one of the six processes to each other to convert into numbers. The linguistic relations obtained were then compared to Table 4 to extract the respective numerical weight. The matrix created for the cost criterion (C) is presented as example (Table 3). This procedure intends to be closer to the original AHP approach due the pairwise comparison, instead of converting the linguistic terms directly into a scale to normalize them.

Table 4: Pairwise relation between the linguistic terms

	1	2	3	4	5	6	7	8	9
	Very very Slow	Very low	Low	Slightly low	Medium	Slightly high	High	very high	very very high
1 Very very slow	1	2	3	4	5	6	7	8	9
2 very low	1/2	1	1 1/2	2	2 1/2	3	3 1/2	4	4 1/2
3 low	1/3	2/3	1	1 1/3	1 2/3	2	2 1/3	2 2/3	3
4 slightly low	1/4	1/2	3/4	1	1 1/4	1 1/2	1 3/4	2	2 1/4
5 Medium	1/5	2/5	3/5	4/5	1	1 1/5	1 2/5	1 3/5	1 4/5
6 slightly high	1/6	1/3	1/2	2/3	5/6	1	1 1/6	1 1/3	1 1/2
7 high	1/7	2/7	3/7	4/7	5/7	6/7	1	1 1/7	1 2/7
8 very high	1/8	1/4	3/8	1/2	5/8	3/4	7/8	1	1 1/8
9 very very high	1/9	2/9	1/3	4/9	5/9	2/3	7/9	8/9	1

The eigenvector obtained from the cost (C) and build time (B) matrixes were employed to build the respective columns to the converted attributes matrix. The numerical values of accuracy (A) and surface roughness (R) were inverted before they were normalized because they are not beneficial values, i. e., lower values are desirable. The values of tensile strength (S) and elongation (E), where higher values are desirable, are simply normalized. Finally, the attributes matrix is built (Table 6).

Table 5: Cost criterion matrix

	Process1	Process2	Process3	Process4	Process5	Process6	Eigenvector
Process1	1,0000	1,0000	0,8750	0,7500	1,0000	0,1250	0,0742
Process2	1,0000	1,0000	0,8750	0,7500	1,0000	0,1250	0,0742
Process3	1,1429	1,1429	1,0000	0,8571	1,1429	0,1429	0,0848
Process4	1,3333	1,3333	1,1667	1,0000	1,3333	0,1667	0,0989
Process5	1,0000	1,0000	0,8750	0,7500	1,0000	0,1250	0,0742
Process6	8,0000	8,0000	7,0000	6,0000	8,0000	1,0000	0,5936
					λmax = 6,0000; CI= 0,0000; CR=0,0000		

Table 6: Attributes matrix to AHP approach

	A	R	S	E	C	B
Process1	0,2053	0,2257	0,3333	0,1235	0,0742	0,1311
Process2	0,1642	0,1174	0,2051	0,2099	0,0742	0,1311
Process3	0,1971	0,0699	0,1538	0,2469	0,0848	0,0820
Process4	0,1331	0,0733	0,1282	0,2469	0,0989	0,1639
Process5	0,2593	0,4191	0,1538	0,1481	0,0742	0,1639
Process6	0,0411	0,0946	0,0256	0,0247	0,5936	0,3279

After evaluating the attributes matrix, the information about criteria (Scenario 1 and Scenario 2) and their weights have to be converted into AHP form. It is therefore necessary, for each scenario, to produce the criteria matrix and to calculate the eigenvector. As an example, a decision maker would define the weights and calculate the eigenvector as presented in Table 7. One should note that the judgments applied to scenario 2 matrix are not consistent, however, the inconsistency is at a low level (CR<0,1) and therefore the matrix may be used.

One should notice that the process capabilities were intentionally not reproduced here. The processes evaluation itself is a hard work, due to constant new development of materials and machines. Best results of process selection are obtained with up-to-date process analysis.

Multiplicative AHP (MAHP)

a) Method

The Multiplicative Analytic Hierarchy Process (MAHP) was developed by Prof. Freeek Lootsma in 1990, and is based on AHP, but uses another scale as well as another algorithm to define the priorities (Eguti et al.,2007). In practice, MAHP has the characteristic to moderate the valuation of "extreme" versus "balanced" alternatives and is less susceptible to rank reversal when adding or removing alternatives (Stam and Silva, 2003).

The MAHP process has the same hierarchy as the AHP. In order to define the relative weight between attributes and criteria, the MAHP uses another scale, as represented in Table 9. As done to AHP, the MAHP requires one matrix for the alternative attributes and n matrixes for the n criteria.

Table 7: Criteria matrix to AHP approach (adapted from [Byun and Lee, 2005])

Scenario 1 – cost of the part (C) and build time (B) considered more important								Scenario 2 – accuracy (A) and surface roughness (R) considered more important						
	A	R	S	E	C	B	eigenvector		A	R	S	E	C	B	eigenvector
A	1	1	3	3	1/5	1/5	0,1113	A	1	1	3	3	5	5	0,3253
R	1	1	3	3	1/5	1/5	0,1113	R	1	1	3	3	5	5	0,3253
S	1/3	1/3	1	1	1/3	1/3	0,0634	S	1/3	1/3	1	1	3	3	0,1113
E	1/3	1/3	1	1	1/3	1/3	0,0634	E	1/3	1/3	1	1	3	3	0,1113
C	5	5	3	3	1	1	0,3253	C	1/5	1/5	1/3	1/3	1	1	0,0634
B	5	5	3	3	1	1	0,3253	B	1/5	1/5	1/3	1/3	1	1	0,0634
λ_{max} = 6,589; CI= 0,118; CR=0,09								λ_{max} = 6,589; CI= 0,118; CR=0,09						

The multiplication of the attributes matrix (Table 6) by the eigenvector of each scenario (Table 7) results in the final ranking.

Table 8: AHP final ranking

scenario 1			scenario 2		
Process	Priority	%	Process	Priority	%
Process6	0,3181	31,81%	Process5	0,2694	26,94%
Process5	0,1721	17,21%	Process1	0,2041	20,41%
Process1	0,1437	14,37%	Process2	0,1508	15,08%
Process4	0,1322	13,22%	Process3	0,1420	14,20%
Process2	0,1244	12,44%	Process4	0,1256	12,56%
Process3	0,1094	10,94%	Process6	0,1082	10,82%

The evaluation of the matrixes is done as explained by Eguti et al.(2007). For each matrix, the weights are transformed into new values, calculated by (4), where δ_{ij} is an integer-valued index designating the decision maker's judgments (Table 9), and γ is a scale parameter. A plausible value for the scale parameter is given by ln 2, which implies on a geometric scale with progression factor 2 (Lootsma, 1996).

$$a_{ij} = e^{\gamma \delta ij} \tag{4}$$

Table 9: Comparison between relative weight scales from AHP to MAHP

Judgements	MAHP (δ_{ij})	AHP (w_i/w_j)
Very strong preference for w_j versus w_i	-8	1/9
Strong preference for w_j versus w_i	-6	1/7
Definite preference for w_j versus w_i	-4	1/5
Weak preference for w_j versus w_i	-2	1/3
Indifference preference for w_i versus w_j	0	1
Weak preference for w_i versus w_j	+2	3
Definite preference for w_i versus w_j	+4	5
Strong preference for w_i versus w_j	+6	7
Very strong preference for w_i versus w_j	+8	9

$$c_i = \frac{1}{n} \sum_{j=1}^{n} a_{ij} \quad i = 1, 2, \ldots n \tag{5}$$

Following, the weights of criteria and attributes matrixes must be calculated. These values are the arithmetical mean, as shown by the equations (5) and (6), respectively.

$$A_{ik} = \frac{1}{m} \sum_{j=1}^{m} a_{ij} \quad i = 1, 2, \ldots m; \quad k = 1, 2, \ldots n \tag{6}$$

$$P_i = \prod_{j=1}^{n} \left(A_{ij} \right)^{c_j} \quad i = 1, 2, \ldots m \tag{7}$$

Where m is the number of alternatives and n the number of criteria. The last step of the MAHP is to obtain the decision vector, using (7).

b) Application

In order to apply the MAHP, the matrixes used for AHP were directly converted using the scale conversion in Table 9 and following the calculations described before. The converted matrixes as well as their respective priority vectors are presented in Table 10.

Table 10: Input matrices for MAHP (Converted from AHP notation)

		scenario 1								scenario 2				
	A	R	S	E	C	B	priority		A	R	S	E	C	B	priority
A	0	0	2	2	-4	-4	0,0914	A	0	0	2	2	4	4	0,3824
R	0	0	2	2	-4	-4	0,0914	R	0	0	2	2	4	4	0,3824
S	-2	-2	0	0	-2	-2	0,0521	S	-2	-2	0	0	2	2	0,0943
E	-2	-2	0	0	-2	-2	0,0521	E	-2	-2	0	0	2	2	0,0943
C	4	4	2	2	0	0	0,3566	C	-4	-4	-2	-2	0	0	0,0233
B	4	4	2	2	0	0	0,3566	B	-4	-4	-2	-2	0	0	0,0233

The calculations of the attributes matrix were carried out as for the AHP. The quantitative attributes had their values inverted (only A and R) and normalized. Relating bothqualitative attributes, one matrix was built to each criterion, in which each alternative was compared to each other, as done for AHP. The conversion from linguistic terms to numerical values was done with a table similar to Table 4, with the respective MAHP values instead of the AHP scale. Following, these matrixes were submitted to the MAHP process to evaluate the local priorities. As a result, the matrix presented here was obtained as the attribute matrix for the MAHP approach.

Table 11: Attribute's matrix of MAHP

	A	R	S	E	C	B
Process1	0,20526	0,22568	0,33333	0,12346	0,00698	0,06985
Process2	0,16421	0,11735	0,20513	0,20988	0,00698	0,06985
Process3	0,19705	0,06985	0,15385	0,24691	0,01405	0,00855
Process4	0,13314	0,07335	0,12821	0,24691	0,02829	0,14066
Process5	0,25928	0,41912	0,15385	0,14815	0,00698	0,14066
Process6	0,04105	0,09464	0,02564	0,02469	0,93674	0,57042

The final evaluation of the MAHP is obtained by multiplying the attribute's matrix by the priority vector of each scenario (Table 10). The results obtained with MAHP for the input data from both previous works are presented and compared in the Table 12.

Table 12: MAHP final ranking

	scenario 1			scenario 2	
Process	Priority	%	Process	Priority	%
Process6	0,5525	55,2%	Process5	0,2914	29,1%
Process5	0,1303	13,0%	Process1	0,2097	21,0%
Process4	0,0986	9,9%	Process2	0,1486	14,9%
Process1	0,0905	9,1%	Process3	0,1404	14,0%
Process2	0,0747	7,5%	Process4	0,1183	11,8%
Process3	0,0533	5,3%	Process6	0,0917	9,2%

VDI Guidelines

The Association of German Engineers (VDI – Verein Deutscher Engenieure) edits regularly guidelines to support engineers to their habitual activities. These guidelines oft support or even become standards. Two VDI guidelines are here considered: The VDI 3404 (2007) and the VDI 2225 (1998).

The VDI 3404 presents, besides definitions regarding layer-manufacturing processes, a simplified method to select processes. It presents generically prototypes criteria and compares them with most significant characteristics of several RP process can offer. The proposed process selection defines some general characteristics of different kind of parts (from visual analysis prototypes up to final products) as well as process properties. However, these definitions are freezed in time. One should consider new process developments offered by additive manufacturing systems suppliers and its own parts requirements.

A pragmatic view of a RP system selection is the assumption that it is a selection procedure inside of the product development process. Pahl et. al. (2006) presented approaches to evaluate decisions during the product development process. Since Rapid Prototyping system selection is a typical application of product development, the guidelines proposed by VDI 2225 are evaluated here.

a) Method

A selection procedure presented by Pahl et. al. (2006) is based on the VDI 2225 (1998), a guideline instruction edited by the Association of German Engineers (VDI). This guideline proposes a simple approach, based on a five-points scale to score the alternatives. The scale and the evaluation table are presented in Table 13Error! Reference source not found.

Table 13: Scale and evaluation table of VDI 2225

Score scale		Technical feature	Alternative A	Alternative t	Ideal
Description	Score	Criterion 1	Wa1	Wt1	4
Very good	4	Criterion 2	Wa2	Wt2	4
Good	3	...	...	...	..
Satisfactory	2	Criterion n	W_{an}	W_{tn}	4
Acceptable	1	Sum	ΣWa	ΣW_t	4.n
Unsatisfactory	0	Technical value x	$\Sigma W_a/4.n$	$\Sigma W_t/4.n$	1
		Economical value y	H_i/H_a	H_i/H_t	1

Where W_{ti} are the scores of the i criterion given to the t alternative following the scale, n the total number of criteria, Hi the ideal manufacturing cost and the Ht the manufacturing cost of the alternative t. Hi can be estimated by $H_i = 0,7.H_{zul}$, where H_{zul} is the permissible manufacturing cost, which is to be determined considering, for example, the lowest price of concurrent products and the revenue margin of the alternative. Some instructions can be found in the literature to predict the cost of each alternative.

VDI 2225 (1998) also considers that the criteria may have different weights. In this case, the technical value should be calculated by (8). Although, VDI do not specify or recommend the scale to weight the alternatives.

$$x = \frac{\sum g_i . w_{ti}}{4. \sum g_i}$$

(8)

Where g_i is the weight of the criterion i.

It is to observe that the computation of costs is done separately by this approach. It is expressed in terms of the economical value y. Further, the VDI 2225 proposes a graphic approach to evaluate the alternative, plotting the technical value x versus the economical value y, defining a point s, in the s-diagram (graph x versus y). VDI suggested that the best solutions have a balanced relationship between cost and technical skills, thus, being nearly the diagonal (traced) line of the s-diagram (Fig. 3).

The s-diagram is also useful to accomplish the evolution of a product. The values s1, s2 and s3 could represent respectively the first, second and third edition of a product. Pahl et. al. (2006) recommends the hyperbole-technique to evaluate the total weight of each alternative, W, by (9).

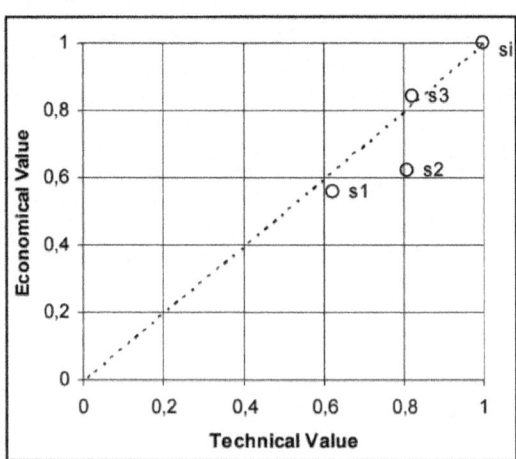

Figure 3: S-diagram example (VDI 2225, 1998)

$$W = \sqrt{(x.y)}$$

$$(9)$$

b) Application

In order to apply the guidelines from VDI 2225, the alternative matrix and criteria matrix have to be converted into the VDI scale and form (Table 13). The conversion table of alternatives to VDI notation is present in Table 14.

Table 14: Alternatives matrix following VDI scale

	A	R	S	E	B
Process1	4	3	4	2	2
Process2	4	2	2	3	2
Process3	4	0	2	4	0
Process4	3	0	1	4	2
Process5	4	4	2	2	2
Process6	0	1	0	0	4

It is to note that attribute Costs (C) were intentionally removed from the Table 14, because VDI proposes a separate economical analysis. The numerical values of the alternatives attributes were mated to the VDI scale, matching the extremity of measured values and of the scale and uniformly distributing the intermediate values. For the attributes A, R and B, the highest values were matched to zero and the lowest to four, because they are unwanted attributes (the higher the value, the less desirable). The calculation of the attributes S and E were made matching the highest values to four, because higher values are desired.

After evaluating the attributes of the alternatives, the following step is to convert the criteria matrixes (the 2 scenarios) to extract the weights used in the VDI guideline. Because the matrixes presented by the previous works are not consistent, it is impossible to extract the exact weight relations among the criteria. Although, in order to compare the different approaches, the following matrices are assumed to be likely representative to the both scenarios (Table 15). One should note that the attribute cost (C) was here also removed.

Table 15: Scenario matrixes into VDI form

	Scenario 1	Scenario 2
A	2	6
R	2	6
S	1	3
E	1	3
B	6	1

The data presented above is enough to perform the calculation of the technical value. The next step consists of calculating the economical value. Once again, some approximations have to be done to allow this estimation, because neither the real cost relation nor the acceptable value is presented.

Since the VDI guideline recommends the economical value to be the relation between the acceptable and the alternative costs, it was considered that the normalized values from the references to be used to represent this relation. VDI also recommends that the acceptable cost should be, if possible, estimated comparing similar products on the market, thus, it was assumed here as the acceptable cost (H_{zul}) being the lowest cost (value) among the normalized alternatives values.

Table 16: VDI 2225 evaluation table for Scenario 2

Criteria	weight (gi)	Process1 Score (wi)	Process1 gi.wi	Process2 wi	Process2 gi.wi	Process3 wi	Process3 gi.wi	Process4 wi	Process4 gi.wi	Process5 wi	Process5 gi.wi	Process6 wi	Process6 gi.wi	Ideal Solution wi	Ideal Solution gi.wi
A	6	4	24	4	24	4	24	3	18	4	24	0	0	4	24
R	6	3	18	2	12	0	0	0	0	4	24	1	6	4	24
S	3	4	12	2	6	2	6	1	3	2	6	0	0	4	12
E	3	2	6	3	9	4	12	4	12	2	6	0	0	4	12
B	1	2	2	2	2	0	0	2	2	2	2	4	4	4	4
Technical value			0,82		0,70		0,55		0,46		0,82		0,13		1
Normalized Cost			1,00		1,00		0,89		0,79		1,00		0,06		
Economical value			0,04		0,04		0,05		0,05		0,04		0,70		

Table 16 presents the results of scenario 2 following the VDI notation. The Fig. 4 represents the evaluation of the W (as (9)) for both scenarios. One should note that due to the separate cost evaluation proposed by VDI, the relative weight of cost compared to the others attributes can not be done. Although, it is to note that the cost has the same weight than all other attributes together, which makes the relative weight of the attributes cost always very high. This can be observed in the Fig. 4, scenario 2, where the accuracy and surface roughness are to be more important, and the process with a lower cost was also the first option. One should notice that due to the separate cost evaluation proposed by VDI, the relative weight of cost compared to the others attributes can not be done. Although, it is also important that the cost has the same weight than all other attributes together, which makes the relative weight of the attributes cost always very high. This can be observed in the Fig. 4, scenario 2, where the accuracy and surface roughness are to be more important, and the process with a lower cost was also the first option.

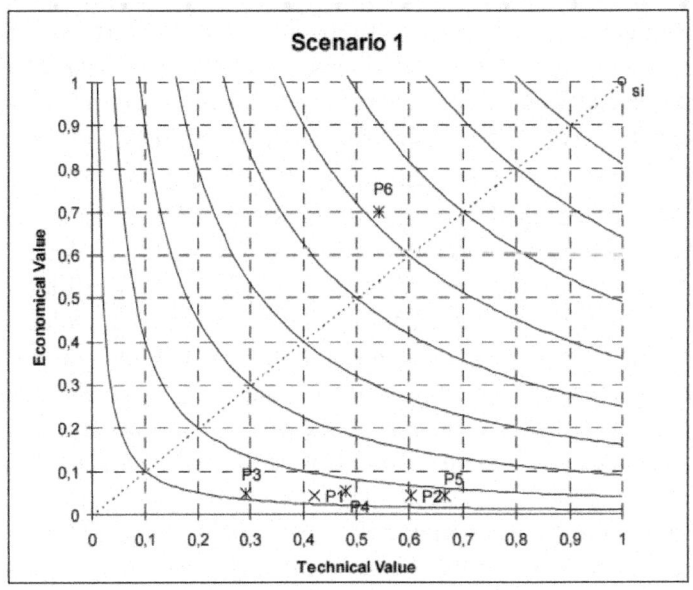

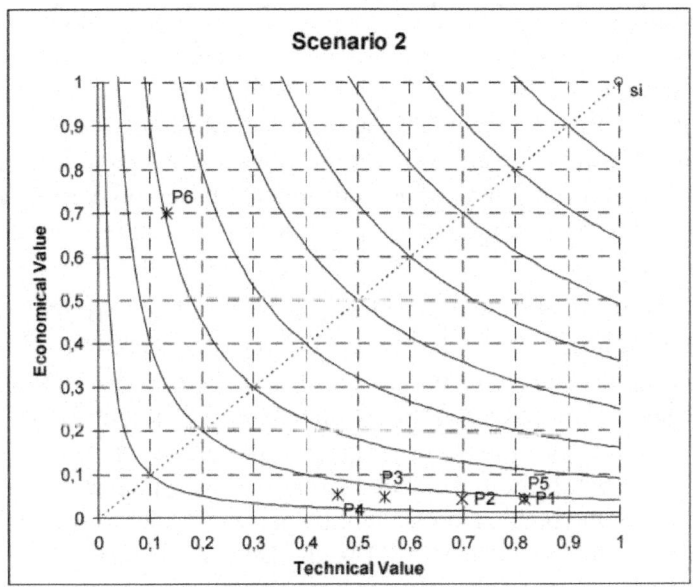

Figure 4: VDI 2225 graphic results for Scenario 1 and Scenario 2.

PART II: RAPID MANUFACTURING APPLICATION

The main advantages of additive manufacturing technologies (AMT) are related to the ability to build geometrically complex shapes without tooling and with high process automation. These characteristics are very useful when producing prototypes, but they can be even more advantageous for final products, if AMT can be integrated into product development. It is because final products may allow the designers and engineers to improve part functionality using more complex shapes. Prototypes have usually a defined form, which may not be modified.

However, some conditions are necessary in order to use AMT for final parts. These conditions are related to lot sizes, shape complexity and costs – AMT are still expensive manufacturing processes. At small lot sizes, such as with customized products, traditional manufacturing technologies become expensive due to high costs of required tooling. Small lot sizes and complex shaped parts are typical features encountered in the aircraft industry. This chapter presents a decision support method based on processes technological information concerning Rapid Manufacturing of plastic parts for aircraft cabin interiors. Nowadays, two RP Technologies are able to process plastic materials, which comply flammability requirements: Fused Deposition Modeling (FDM) and Selective Laser Sintering (SLS). A method is presented to consider the possible advantages and restrictions when considering the manufacturing process. Further, a procedure to evaluate quality, productiontime and cost is presented. The method is illustrated with examples on the selection of manufacturing technology to produce a customized decoration part and an air duct. Typical costs and manufacturing time of injection moulding processes were also compared and analyzed with the proposed method. It is possible to define the break-even point, when conventional processes become preferred then AMT. Fig. 5 illustrates the general process selection presented in this work.

Fig. 5 presents also the parallel comparison with a conventional process chain. Since all parts are so far designed to be produced by processes other then AMT (called here conventional processes), there is always an alternative process chain. It has, in turn, been optimized over years, and the costs, quality and delivery time quite known by manufacturing engineers. It is not the aim of this work to select the conventional alternative, but, typical delivery time and costs related to the both examples will be presented later in this chapter. The proposed procedure to evaluate AMT is divided into two phases: 1) analysis of requirements; and 2) classification and prioritization, as explained in the next sections.

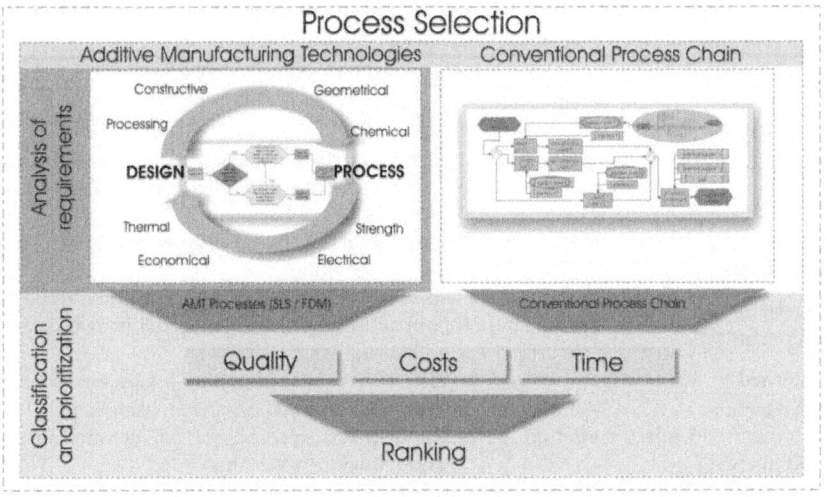

Figure 5: Material restriction when considering RT in the aircraft industry.

Analysis of Requirements

Analysis of requirements aims to eliminate processes – or process chains – which do not provide adequate properties. It begins with the material analysis. Grimm (2004) argues that material selection may lead to a manageable quantity of process to analyze. Thus, it should be performed first.

The Association of German Engineers (VDI), in the outline of guideline VDI 3404 (2007), presents generically parts requirements (Table 17). Decision makers should use it as check list when summarizing their parts requirement. The quality of a part is also related to how its function is performed. Thus, AMT must assure these requirements. Each specificrequirement should be analyzed based on process information (process attributes) found in literature, but even more important, based on up-to-date analyses. They could be obtained directly at manufacturers and resellers, but they are usually not specific enough. The tensile strength is an example, where the manufacturer information does not specify the material resistance among different building directions. Alternatively, attributes or rule databases (Masood and Soo, 2002; Katschka, 1999) could be used, but with restrictions. Furthermore, a large amount of work would be needed to maintain such databases up to date. The process attributes used in this work were available in the literature (Borille, 2009).

Table 17: Quality characteristics of part requirements (adapted from VDI3404, 2007)

Requirements	Relevant quality characteristics
Constructive requirements	Size, scale, weight, density, textures, colors / transparency, odor
Geometrical requirements	Component size and complexity, length and angle dimensions, dimensional tolerances, form and position deviations, shrinkage, minimal structures, walls, layer thicknesses
Processing requirements	Machinability, formability, joinability, Surface finishing (painting, coating, polishing)
Strength requirements	Tensile, compression, bending and torsion strength, static and dynamic creep rupture strength, impact strength, hardness, friction coefficient, abrasion
Thermal requirements	Use temperature ranges, resistance to heat, softening temperature, specific heat, thermal conductivity, thermal expansion coefficient
Electrical requirements	Dielectric strength, surface and spec. Contact resistance, dielectric property values, tracking resistance
Chemical requirements	Flammability, toxicity, resistance to aggressive media, water absorption, biocompatibility, light stability, light transmission
Economical requirements	Units/lot size, production times/delivery times, production costs, reliability, waste and disposal costs

In order to evaluate the requirements, the logical question associated to each one is if process and/or material meet the requirement. However, there are two further questions proposed: 1) if the requirement is not met is it possible to meet the requirement by means of design modifications? 2) Is it possible to improve the part quality or reduce cost by means of design modifications? Fig, 6 presents the sequential decision regarding the verification of a requirement.

This verification aims at inducing the decision maker to think about all the possibilities regarding AMT. Freedom of form and process flexibility should be always in mind when answering these questions. The potential of implementing AMT lies on the component improvement, which can be as weight reduction, reduction of parts quantity by assembling components, reducing costs of complex shapes among others.

Economical requirements, expressed by the cost, have two major functions in the proposed methods. First, in the initial procedure phase, the cost should be use as a filter to eliminate alternatives which are not at reasonable levels. The cost of each alternative will be needed later again, when creating the alternatives ranking, comparing with their quality and fabrication time. It should be interesting to create a database containing the considerationsof each requirement (Design solutions). Applied design solutions could be based on the results from previous processes.

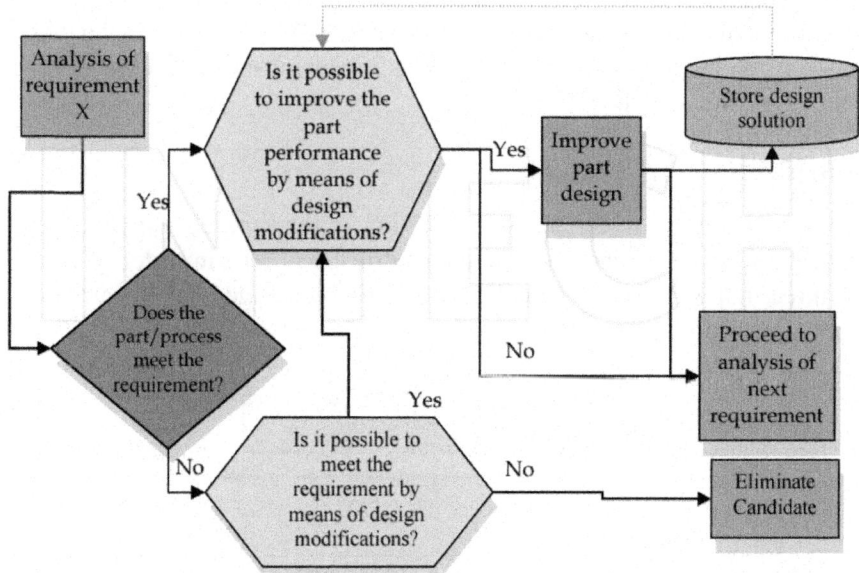

Figure 6: Analysis of requirements.

This procedure aims to evaluate whether an AMT process is able to provide adequate technical parts. It is a filtering procedure, but, it also aims to integrate product and process. There is a reason why not to classify the process (create a ranking) at this phase, as proposed in the literature (Rao and Padmanabhan, 2007; Rao, 2007). It is because the technical analysis is done separately from cost and time. Cost and time are usually associated to – low values, better values -, but most technical requirements can not be analyzed this way. It is difficult to argue that a part, which present surface roughness $R_a = 2$ μm, is five times better than other which has 10 μm, when the specification is 15 μm. It is correct to affirm that the both processes are good enough regarding this requirement. Even when scale normalization is used, the rates between requirements could still carry such inconsistencies.

Technical requirements act as filters, but they also carry information for the second phase. All the technical considerations should be stored under – Quality – and will be used to generate the final ranking – Classification and prioritization. Each relevant aspect observed when considering the requirements should be aggregated within – Quality.

In doing the analysis of requirements before observing the costs, it is expected that all improvement possibilities are checked and aggregated together. If the part improvement reaches a high level, it can be strong enough to be contrasted to cost. One frequent characteristic observed in industries when studying the possibilities to apply AMT, is thecost evaluation as first consideration. As the material costs are comparatively expensive, the technology is rejected.

Classification and Prioritization

The second phase of the process selection is the ranking generation. It is proposed to use the Analytic Hierarchy Process (AHP) in order to evaluate the three major aspects: quality, cost and time. The general hierarchy of is expresses as in Fig. 7.

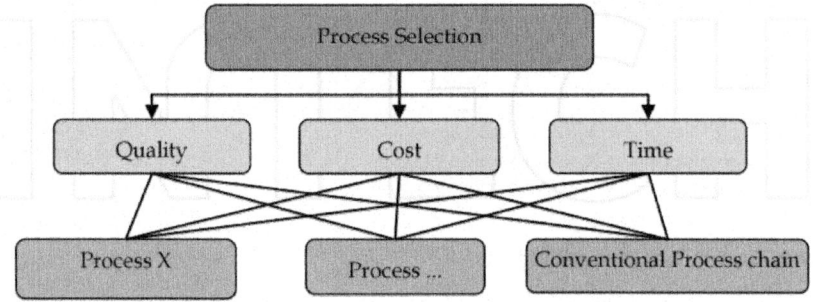

Figure 7: The proposed hierarchical structure of AMT process selection.

In the following sections this procedure is applied to two case studies as application examples.

Example of Application

The parts analyzed in the context of this work are presented in Fig. 8.

Part 1 – Air duct

Part 2 – Decoration part

Figure 8: Representative parts.

The first part consists of an air duct. The main features are associated to the complex shape and the usual need for assemblies and fixture elements, which were integrated in the design. Part two represents a customized panel, which could include logos, as represented. Esthetical aspects and flexibility to produce different forms at low lot sizes represent great importance to consider the manufacturing process. Air duct is a typical example of AMT in the aircraft industry (DeGrange, 2006; Hopkinson et. al., 2006; Aerospace Engineering, 2004).

The part was modelled including features which are not usually integrated, as fixture elements, one-piece-body and internal walls to direct air flow. Some part requirements are presented in Table 18.

Table 18: Air duct requirements

Requirements	
Constructive	Max. dimensions: 69; 204; 160 mm
Strength	Good properties in all directions
Geometrical	Duct with curvature in two directions, wall thickness 1,5 mm, max. form deviation 0,5mm/100 mm.
Processing	Coating and sealing required
Chemical	Flammability,

After applying the verification procedure described in Fig, 6, it was observed that both FDM and SLS processes meet the requirements. In order to enable support structures removal the part produced by FDM had to be correctly positioned related to the build up direction. SLS enable also the integration of additional functions compared to FDM, exemplified by the introduction of a diffusor at one extremity. This part was produced by both processes, FDM and SLS, as Fig. 9.

Due to the support structures need, the FDM building process was restricted to one build up direction. This part positioning related to the layers was selected to avoid deposition of support material in regions where its removal could not be done. The satisfied product requirements in Table 18 are not used anymore, but the relevant quality aspects, which are aggregated in Table 19. These aspects have to be in mind to the next phase of selection procedure. Relevant aspects are related to requirements which can be performed more efficiently by using AMT resulting in desired part improvements. Requirements as accuracy specified as being less then a certain value usually do not improve product quality. They should be considered as a filter to eliminate inadequate processes. However, higher tensile strength materials may be used to reduce weight, which may be a product improvement. Evidently, if one process can not satisfy one or more requirements, it should be excluded form the selection process.

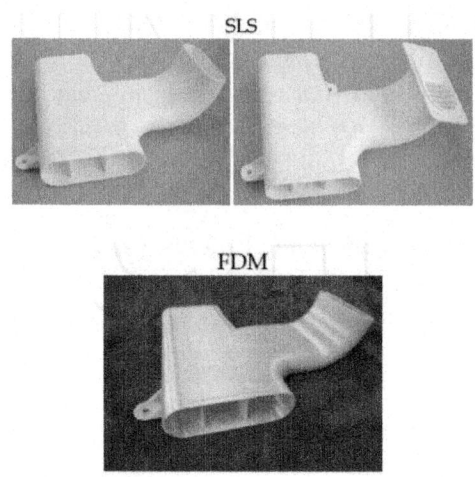

Figure 9: AMT manufactured air ducts.

Table 19: Aggregated process attributes for air duct part

Requirements	SLS	FDM
Constructive	Better form flexibility because no support structures are required	Restrictions due to support structures
Strength	Better isotropic material behavior	
Geometrical	Duct with curvature in two directions, wall thickness 1,5 mm	

As described, the second process selection method phase consists on creating the rank based on weightening quality, cost and time according user needs. Typical applications require low cost. Sometimes the time may be more important or even the quality. In order to exemplify, the next estimations are presented as cost preference, it means that cost is preferred instead of delivery time and quality. How much cost is preferred will be defined using requirements prioritization within AHP method.

Cost Preference

Considering three alternatives and three requirements, four matrices should be filled with pairweise comparisons. The first one refers to comparison among the requirements to identify their priorities. Following, all the alternatives have to be compared considering each requirement. As this example has three major requirements (cost, time and quality), three additional matrices are required.

The decision team should fill these matrices with judgments according the fundamental scale of Saaty (Saaty, 2000), presented in Table 2. As quantitative requirements are presented (cost and delivery time), it is possible to fill the matrices with their rates instead of Saaty's fundamental scale. In this case, one should take care to notice whether the desired values are the higher or the lower ones.

The priority related to each matrix is represented by its eigenvector, thus, they have to be calculated to all matrices. A matrix is built assembling the resulting eigenvectors from the alternatives comparison matrices. This resulting matrix, in turn, is then multiplied by the eigenvector resulting from the requirements comparison table. This example considers the costs as being stronger than other requirements. As possible judgments, it was considered that cost is strongly preferred than quality and time delivery, and quality slightly then time. These judgments have to be translated into a matrix, represented in Table 20.

Table 20: Requirements comparison matrix

	Cost	Quality	Time	Eigenvector
Quality	1/7	1	2	0,1392
Cost	1	7	7	0,7732
Time	1/7	1/2	1	0,0877
	λ_{max}= 3,0536; CI= 0,0268; CR= 0,0516			

The eigenvector presented in Table 20 represents a numerical ranking of the requirements. It translates the decision team preferences into numerical values. The ordering also reflects intensity as indicated by the ratios of the numerical values. It is worth noticing that the AHP allows certain inconsistencies, which are represented by the CR. CR values less than 10% (0,1) are considered acceptable (Saaty, 1977). If CR is greater than 10%, the judgments have to be revised. In this example, the inconsistency relies on the fact that cost has the same importance rate to quality and time, however quality is judged more important then time. The next step consists on compare the alternatives considering each requirement. At this point, quality is represented by extra functionality which may be performed using AMT, according Table 19. Time and cost were analyzed in the reference (Borille, 2009). Table 21 represents the judgments related to quality of the processes. As SLS allows the integration of additional functions, it is considered more important than FDM. FDM in turn, makes it possible the integration of fixture elements when compared to conventional processes, thus, being also more important then conventional processes.

Table 21: Alternative matrix for requirement quality

Quality	SLS	FDM	Conventional	Eigenvector
SLS	1	3	7	0,6694
FDM	1/3	1	3	0,2426
Conventional	1/7	1/3	1	0,0879
	λ_{max} = 3,0070; CI= 0,0035; CR=0,0068			

Regarding cost and time, this example uses the values obtained from service provides. Different service providers offer different prices and delivery times. The costs are resumed in Table 22. Time is considered as being the delivery time of the first produced part. As cost is preferred, the less expensive alternatives were selected.

Table 22: Cost and time for purchasing the air duct part

	AMT		Conventional
	SLS	FDM	
Cost [RS$]	1.674,40	1.371,06	*Variable according number of parts*
Time [business days]	3	7	20

Conventional process costs per unit are strongly dependent on quantity of produced parts. Costs estimation will be used to define the minimal batch size, which conventional process becomes preferable then AMT. This number is called Break-even-point (Zäh, 2006). Table 23 represents the team's judgments regarding time. Table 24 exemplifies the judgments for requirement cost. As quantitative values are available, they are used instead of building another comparison matrix. The costs are normalized and their inverse values are used because lower costs are desired. In order to simulate diverse batch sizes, Table 24 was reproduced using different values of conventional process cost per unit.

Table 23: Alternative matrix for requirement time

Time	SLS	FDM	Conventional	Eigenvector
SLS	1	3	5	0,6370
FDM	1/3	1	3	0,2583
Conventional	1/5	1/3	1	0,1047
λ_{max} = 3,0385; CI= 0,0193; CR=0,0370				

Table 24: Alternative matrix for requirement cost – example for 10 parts

	Cost [RS$]	Preference
SLS	1.674,40	0,3919
FDM	1.371,06	0,4786
Conventional process (10 parts)	5.069,69	0,1294

The final ranking results from multiplying the matrices presented in Table 25. In this case, for ten parts, SLS process would be selected with 45% of preference, although FDM is the cheapest alternative.

Table 25: Final ranking generation (10 parts)

	Quality	Cost	Time	Requirements	Ranking
SLS	0,6694	0,3919	0,6370	0,1392	0,4521
FDM	0,2426	0,4786	0,2583	0,7732	0,4265
Conventional	0,0879	0,1294	0,1047	0,0877	0,1215

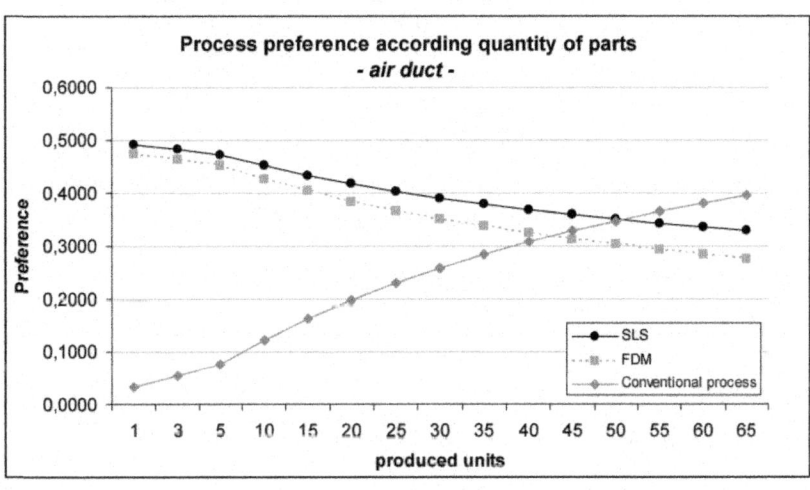

Figure 10: Simulation of process preference quantity of produced parts for part 1.

Varying the quantity of produced parts, conventional injection molding processbecomes preferred because the cost per part decreases significantly. Using the proposed procedure, one can estimate the break-even-point. Fig.

10 shows that, in this case, SLS process would be preferred until batch sizes of approximately 50 parts. Larger batches should be produced using injection molding. When AMT batch size becomes larger, it should be considered that the produced parts delivery time may increase depending on the machine capacity of the service provider. The price per part, in this case, may also be reduced due to the better machine usage, specially when considering SLS (Borille, 2009).

Case Two: Interior Decoration Part

The same selection procedure was applied to the part two, an example of decoration part. The quality attributes are aggregated in Table 26, which presented also the manufactured parts.

Table 26: Aggregated process attributes for decoration part

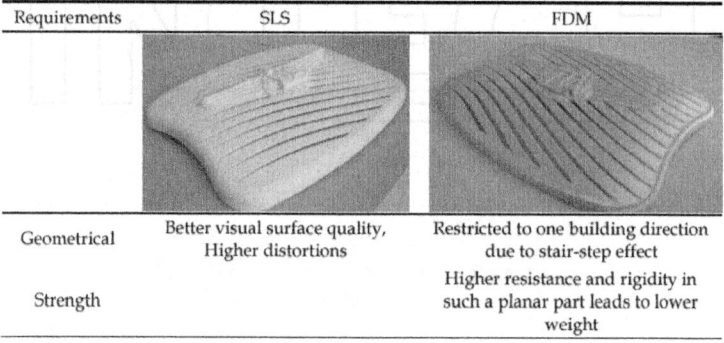

Requirements	SLS	FDM
Geometrical	Better visual surface quality, Higher distortions	Restricted to one building direction due to stair-step effect
Strength		Higher resistance and rigidity in such a planar part leads to lower weight

The decision team faced the following situation: the customers needs consist on the quickly customization of its aircraft. As requirements, the decision team built up the following requirements matrix, Table 27.

Table 27: Requirements comparison matrix for decoration part

	Quality	Cost	Time	Eigenvector
Quality	1	3	1/3	0,2308
Cost	1/3	1	1/9	0,0769
Time	3	9	1	0,6923
λ_{max}= 3,0000; CI= 0,0000; CR= 0,0000				

Table 28 presents the decision team judgments for quality, according considerations from Table 26. Although the better surface quality of SLS, the FDM process may produce stronger planar parts due to its higher tensile resistance. SLS and FDM are considered as the same importance. Injection molding process presents some restrictions due to draft angles to allow the mold opening, thus, it was considered less important.

Table 28: Alternative matrix for requirement quality

Quality	SLS	FDM	Conventional	Eigenvector
SLS	1	3	3	0,4286
FDM	1	1	3	0,4286
Conventional	1/3	1/3	1	0,1429
	λ_{max} = 3,0000; CI= 0,0000; CR=0,0000			

When purchasing the fastest alternatives from service providers, the cost values arc used to judge the alternatives regarding time in Table 29 and to build the cost rates in Table 30.

Table 29: Alternative matrix for requirement time

Time	SLS	FDM	Conventional	Eigenvector
SLS	1	2	9	0,5969
FDM	1/2	1	7	0,3458
Conventional	1/9	1/7	1	0,0572
	λ_{max} = 3,0217; CI= 0,0109; CR=0,0209			

Table 30: Alternative matrix for requirement cost – example for 5 parts

	Cost [RS$]	Preference
SLS	1034,80	0,6025
FDM	1040,00	0,3284
Conventional process (5 parts)	4945,44	0,0691

The final ranking results from multiplying the matrices presented in Table 31. In this case, for five parts, SLS process would be selected with 55% of preference.

Table 31: Final ranking generation – decoration part

	Quality	Cost	Time	Requirements	Ranking
SLS	0,4286	0,6025	0,5969	0,2308	0,5585
FDM	0,4286	0,3284	0,3458	0,0769	0,3636
Conventional	0,1429	0,0691	0,0572	0,6923	0,0779

SLS process was selected because it had in this example the lower price and the shorter delivery time. In this example, the cost per part reduction of injection molding could not overcome the time requirement.

Fig. 11 represents the preference ranking. Injection molding would be effective only when the parts quantity become high enough to imply in higher SLS delivery time.

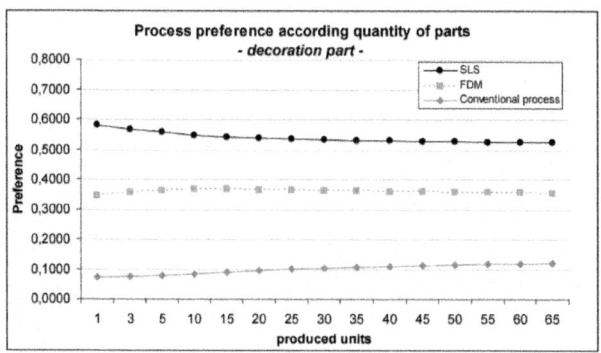

Figure 11: Simulation of process preference quantity of produced parts for part 2.

CONCLUSION

Rapid Manufacturing is becoming reality in several industries, among them the aeronautical. New machine and the further material developments allow the continuous expansion of applications. Grimm (2004) mentioned that there was at that time no machine with focus on RM. Three years later, Arcam presented the machine called A2, which is considered the first one focused on RM applications (Arcam, 2007). Further examples of these trends were presented at the Euromold 2008 trade fair, in Frankfurt, Germany. Stratasys as well as EOS presented new material options and new machines. Ultem© for FDM equipments and PEEK for SLS are both high performance polymers and potential candidates to be used in aircraft applications by means of AMT. The introducing into the market of both new materials choices as well new machine generations are important indicators of the aircraft industry market importance. However, the method suggested in this work could be applied not only for aeronautical applications. It could also more options to compare and choose the best alternative considering also the new alternatives.

Another point which would contribute to the implementation of this procedure is the definition of metrics to aggregate components according geometrical similarity. The presented work was based on visual similarities to select models as representative geometries and proposed the individual cost and build time estimation. But users could develop definitions of metrics which could represent groups of parts. It could accelerate the cost and time estimation. Make or buy decision could also be done based on results from the proposed procedure. The point to be analysed is the estimation of quantity of parts that the company would like to produce. This quantity should be used to calculate the machine cost per hour, which is one of major cost factor.

ACKNOWLEDGMENT

The authors would like to thank especially Prof. Dr.-Ing. Karl-Heinrich Grote at Otto-vonGuericke University and Dr. Rudolf Meyer at Fraunhofer IFF in Magdeburg and also Prof. Dr.-Ing. Fritz Klocke at RWTH and Dipl.-Ing. Axel Demmer at Fraunhofer IPT in Aachen, Last but not least, this work could not have been completed without the financial support from FAPESP (Fundação de Amparo à Pesquisa do Estado de São Paulo), FINEP (Financiadora de Estudos e Projetos), CNPq (Conselho Nacional de Desenvolvimento Científico e Tecnológico), CAPES (Coordenação de Aperfeiçoamento de Pessoal de Nível Superior) and DAAD (Deutscher Akademischer Austausch Dienst).

REFERENCES

1. Borille, A. V. Decision support method to apply Additive Manufacturing Technologies for plastic components in the aircraft industry. Thesis of doctor in science – Program of Mechanics Engineering, area of Aerospace Systems and Mechatronics. Technological Institute of Aeronautics. São José dos Campos, 2009.

2. Borille, A. V., Gomes, J. O., Grote, K.-H., Meyer, R. The use of decision methods to select Rapid Prototyping technologies, Rapid Prototyping Journal. Paper approved for publishing in issue 1, Vol 16, 2010

3. Byun, H. S., Lee, K. H. (2005). A decision support system for the selection of rapid prototyping process using the modified TOPSIS method. International Journal of Advanced Manufacturing Technology, Vol 26, pg 1338-1347.

4. Katschka, U. Methodik zur Entscheidungsunterstützung bei der Auswahl und Bewertung von Konventionellen und Rapid Tooling-Prozessketten. PhD thesis. Technische Universität Chemnitz. Shaker Verlag. ISBN 3-8265-6431-6, 1999.

5. Rao, R. V., Padmanabhan, K. K. (2007). Rapid Prototyping process selection using graph theory and matrix approach. Journal of Materials Processing Technolog, Volume 194, Issues 1-3, 1 November, Pages 81-88.

6. Saaty, T. L. (2003) Decisiont-making with the AHP: Why is the principal eigenvector necessary, European Journal of Operational Research, Vol. 145, pp. 85-91.

7. VDI 2225 (1998). Konstruktionsmethodik: Technisch-witschaftliches Konstruiren, technischwitschaftliche Bewertung. Verein Deutscher Ingenieure, November.

8. VDI 3404 (2007). Generative Fertigungsverfahren: Rapid-Technologiens (Rapid Prototyping) Grundlagen, Begriffe, Qualitätskenngrößen, Liefervereinbarungen. Verein Deutscher Ingenieure, December.

9. Zäh, Michael F. (2006). Wirtschaftliche Fertigung mit Rapid-Technologien. Anwender-Leitfaden zur Auswahl geeigneter Verfahren. Hanser Verlag. ISBN 3-446-22854-3

CITATION

CHAPTER 1

D. Dolage and A. Sade, "A Frontier Approach to Measuring Impact of Adoption of Flexible Manufacturing Technology on Technical Efficiency of Malaysian Manufacturing Industry," Technology and Investment, Vol. 3 No. 4, 2012, pp. 266-275. doi: 10.4236/ti.2012.34037.

CHAPTER 2

Keith Swain, DuPont, Wilmington, Delaware; Nanomaterial Production and Downstream Handling Processes; http://www.cdc.gov/niosh/docs/2014-102/pdfs/2014-102.pdf

CHAPTER 3

Soosung Kim, Kihwan Kim, Jungwon Lee, and Jinhyun Koh, "Design and Fabrication of Remote Welding Equipment in a Hot-Cell," Science and Technology of Nuclear Installations, vol. 2013, Article ID 970942, 8 pages, 2013. doi:10.1155/2013/970942

CHAPTER 4

KonstantinChuntonov,JanezSetina,GaryDouglass, (2015) The Newest Getter Technologies: Materials, Processes, Equipment. Journal of Materials Science and Chemical Engineering,03,57-67. doi: 10.4236/msce.2015.39008

CHAPTER 5

V. Kondratenko, V. Borisovsky, A. Naumov and N. Petruljanis, "New Technology for Grids and Scales Manufacturing in Optical Devices," Optics and Photonics Journal, Vol. 2 No. 3, 2012, pp. 163-166. doi:10.4236/opj.2012.23024.

CHAPTER 6

Mantrala KM, Das M, Balla VK, Rao CS and Kesava Rao VVS. (2015) Additive manufacturing of Co-Cr-Mo alloy: influence of heat treatment on microstructure, tribological, and electrochemical properties. Front. Mech. Eng. 1:2. doi: 10.3389/fmech.2015.00002

CHAPTER 7

Hasan Hosseini-Nasab, Mohammad Dehghani, and Amin Hosseini-Nasab, "Analysis of Technology Effectiveness of Lean Manufacturing Using System Dynamics," ISRN Industrial Engineering, vol. 2013, Article ID 237402, 10 pages, 2013. doi:10.1155/2013/237402

CHAPTER 8

J. Li, J.Y. H. Fuh, Y.F. Zhang and A.Y.C. Nee (2006). Multi-Agent Based Distributed Manufacturing, Manufacturing the Future, Vedran Kordic, Aleksandar Lazinica and Munir Merdan (Ed.), ISBN: 3-86611-198-3, InTech, DOI: 10.5772/5043,

CHAPTER 9

Vittorio Cesarotti, Alessio Giuiusa and Vito Introna (2013). Using Overall Equipment Effectiveness for Manufacturing System Design, Operations Management, Prof. Massimiliano Schiraldi (Ed.), ISBN: 978-953-51-1013-2, InTech, DOI: 10.5772/56089.

CHAPTER 10

Tauseef Aized (2010). Material Handling in Flexible Manufacturing System, Future Manufacturing Systems, Tauseef Aized (Ed.), ISBN: 978-953-307-128-2, InTech, DOI: 10.5772/10241.

CHAPTER 11

Tritos Laosirihongthong (2006). Multidimensional of Manufacturing Technology, Organizational Characteristics, and Performance, Manufacturing the Future, Vedran Kordic, Aleksandar Lazinica and Munir Merdan (Ed.), ISBN: 3-86611-198-3, InTech, DOI: 10.5772/5067.

CHAPTER 12

Anderson Vicente Borille and Jefferson de Oliveira Gomes (2011). Selection of Additive Manufacturing Technologies Using Decision Methods, Rapid Prototyping Technology - Principles and Functional Requirements, Dr. M. Hoque (Ed.), ISBN: 978-953-307-970-7, InTech, DOI: 10.5772/24045.

INDEX